5 MOST DEADLIEST DISEASES

VOLUME 1

Author

Arin Bhattacharya

DEDICATION

This book is dedicated

to

My respected parents

I also thank god, my fellow mates

&

All the lovers of pharmacy world

Acknowledgement

In any of scientific literally work many people work towards achievement of the desired goal i.e. publication of the research work in form book. I want to thanks my parents without *their blessings, support and vision I could not have reached this stage.*

Friendship is a treasured gift and fine friends are very few, My heartful thanks to my friend s who are always there for their support and positive views. I also want to thank someone very special on FB, whose inspiration played a very important role in completing this work and without her this work would not be possible.

I would like to thanks the persons whom had criticize this project because it provided the positive impetus for completing the project in a more comprehensive and positive way.

Lastly it is only when someone writes a book that one realizes the true power of MSOffice software. It is simple- without this software this book would not be written.

Thank you Mr. Bill Gates and Microsoft Corp!

I strongly believe, "Success is blend of hard work and Destiny" and I think 'God' who have perpetually patronized me with consciousness and love to ladder the success. Thankful I ever remain....

Arin Bhattacharya

Table of Contents

CHAPTER 1 WHAT IS DISEASE

A disease is an abnormal condition that affects the body of an organism. It is often construed as a medical condition associated with specific symptoms and signs It may be caused by factors originally from an external source, such as infectious disease, or it may be caused by internal dysfunctions, such as autoimmune diseases. In humans, "disease" is often used more broadly to refer to any condition that causes pain, dysfunction ,distress, social problems, or death to the person afflicted, or similar problems for those in contact with the person. In this broader sense, it sometimes includes injuries, disabilities, disorders, syndromes, infections, isolated symptoms, deviant behaviours, and atypical variations of structure and function, while in other contexts and for other purposes these may be considered distinguishable categories. Diseases usually affect people not only physically, but also emotionally, as contracting and living with many diseases can alter one's perspective on life, and one's personality.

Death due to disease is called death by natural causes. There are four main types of disease: pathogenic disease, deficiency disease, hereditary disease, and physiological disease. Diseases can also be classified as communicable and non-communicable disease.

Concepts

In many cases, the terms *disease*, *disorder*, *morbidity* and *illness* are used interchangeably. In some situations, specific terms are considered preferable.

Disease

The term *disease* broadly refers to any condition that impairs normal function, and is therefore associated with dysfunction of normal homeostasis Commonly, the term *disease* is used to refer specifically to infectious diseases, which are clinically evident diseases that result from the presence of pathogenic microbial agents, including viruses, bacteria, fungi, protozoa, multicellular organisms, and aberrant proteins known as prions. An infection that does not and will not produce clinically evident impairment of normal functioning, such as the presence of the normal bacteria and yeasts in the gut, or of a passenger virus, is not considered a disease. By contrast, an infection that is asymptomatic during its incubation period, but expected to produce symptoms later, is usually considered a disease. Non-infectious diseases are all other diseases, including most forms of cancer, heart disease, and genetic disease.

Illness

Illness and *sickness* are generally used as synonyms for *disease*.[1] However, this term is occasionally used to refer specifically to the patient's personal experience of his or her disease

In this model, it is possible for a person to have a disease without being ill (to have an objectively definable, but asymptomatic, medical condition), and to be *ill* without being *diseased* (such as when a person perceives a normal experience as a medical condition, or medicalizes a non-disease situation in his or her life). Illness is often not due to infection, but to a collection of evolved responses—sickness behavior by the body—that helps clear infection. Such aspects of illness can include lethargy, depression, anorexia, sleepiness, hyperalgesia, and inability to concentrate.

Disorder

In medicine, a disorder is a functional abnormality or disturbance. Medical disorders can be categorized into mental disorders, physical disorders, genetic disorders, emotional and behavioral disorders, and functional disorders. The term *disorder* is often considered more value-neutral and less stigmatizing than the terms *disease* or *illness*, and therefore is preferred terminology in some circumstances. In mental health, the term mental disorder is used as a way of acknowledging the complex interaction of biological, social, and psychological factors in psychiatric conditions. However, the term *disorder* is also used in many other areas of medicine, primarily to identify physical disorders that are not caused by infectious organisms, such as metabolic disorders.

Medical condition

A medical condition is a broad term that includes all diseases and disorders. While the term *medical condition* generally includes mental illnesses, in some contexts the term is used specifically to denote any illness, injury, or disease except for mental illnesses. The Diagnostic and Statistical Manual of Mental Disorders (DSM), the widely used psychiatric manual that defines all mental disorders, uses the term *general medical condition* to refer to all diseases, illnesses, and injuries except for mental disorders This usage is also commonly seen in the psychiatric literature. Some health insurance policies also define a *medical condition* as any illness, injury, or disease except for psychiatric illnesses

As it is more value-neutral than terms like *disease*, the term *medical condition* is sometimes preferred by people with health issues that they do not consider deleterious. On the other hand, by emphasizing the medical nature of the condition, this term is sometimes rejected, such as by proponents of the autism rights movement.

The term *medical condition* is also a synonym for *medical state*, in which case it describes an individual patient's current state from a medical standpoint. This usage appears in statements that describe a patient as being *in critical condition*, for example.

Morbidity

Morbidity (from Latin *morbidus*, meaning "sick, unhealthy") is a diseased state, disability, or poor health due to any cause The term may be used to refer to the existence of any form of disease, or to the degree that the health condition affects the patient. Among severely ill patients, the level of morbidity is often measured by ICU scoring systems. Comorbidity is the simultaneous presence of two or more medical conditions, such as schizophrenia and substance abuse.

In epidemiology and actuarial science, the term "morbidity rate" can refer to either the incidence rate, or the prevalence of a disease or medical condition. This measure of sickness is contrasted with the mortality rate of a condition, which is the proportion of people dying during a given time interval.

Syndrome

A syndrome is the association of several medical signs, symptoms, and or other characteristics that often occur together. Some syndromes, such as Down syndrome, have only one cause; others, such as Parkinsonian syndrome, have multiple possible causes. In other cases, the cause of the syndrome is unknown. A familiar syndrome name often remains in use even after an underlying cause has been found, or when there are a number of different possible primary causes.

Predisease

Predisease is a type of disease creep or medicalization in which currently healthy people with risk factors for disease, but no evidence of actual disease, are told that they are sick . Prediabetes and prehypertension are common examples. Labeling a healthy person with predisease can result in overtreatment, such as taking drugs that only help people with severe disease, or in useful preventive measures, such as motivating the person to get a healthful amount of physical exercise

Types

Infectious diseases

Contagious diseases

Foodborne illness

Foodborne illness or food poisoning is any illness resulting from the consumption of food contaminated with pathogenic bacteria, toxins, viruses, prions or parasites.

Communicable diseases

Non-communicable diseases

Airborne diseases

Lifestyle diseases

A lifestyle disease is any disease that appears to increase in frequency as countries become more industrialized and people live longer, especially if the risk factors include behavioral choices like a sedentary lifestyle or a diet high in unhealthful foods such as refined carbohydrates, trans fats, or alcoholic beverages.

Mental disorders

Mental illness is a broad, generic label for a category of illnesses that may include affective or emotional instability, behavioral dysregulation, and/or cognitive dysfunction or impairment. Specific illnesses known as mental illnesses include major depression, generalized anxiety disorder, schizophrenia, and attention deficit hyperactivity disorder, to name a few. Mental illness can be of biological (e.g., anatomical, chemical, or genetic) or psychological (e.g., trauma or conflict) origin. It can impair the affected person's ability to work or study and harm interpersonal relationships. The term insanity is used technically as a legal term.

Organic disease

An organic disease is one caused by a physical or physiological change to some tissue or organ of the body. The term sometimes excludes infections. It is commonly used in contrast with mental disorders. It includes emotional and behavioral disorders if they are due to changes to the physical structures or functioning of the body, such as after a stroke or a traumatic brain injury, but not if they are due to psychosocial issues.

Stages

In an infectious disease, the incubation period is the time between infection and the appearance of symptoms. The latency period is the time between infection and the ability of the disease to spread to another person, which may precede, follow, or be simultaneous with the appearance of symptoms. Some viruses also exhibit a dormant phase, called viral latency, in which the virus hides in the body in an inactive state. For example, varicella zoster virus causes chickenpox in the acute phase; after recovery from chickenpox, the virus may remain dormant in nerve cells for many years, and later cause herpes zoster (shingles).

Acute disease

An acute disease is a short-lived disease, like the common cold.

Chronic disease

A chronic disease is one that lasts for a long time, usually at least six months. During that time, it may be constantly present, or it may go into remission and periodically relapse. A chronic disease may be stable (does not get any worse) or it may be progressive (gets worse over time). Some chronic diseases can be permanently cured. Most chronic diseases can be beneficially treated, even if they cannot be permanently cured.

Flare-up

A flare-up can refer to either the recurrence of symptoms or an onset of more severe symptoms.

Refractory disease

A refractory disease is a disease that resists treatment, especially an individual case that resists treatment more than is normal for the specific disease in question.

Progressive disease

Progressive disease is a disease whose typical natural course is the worsening of the disease until death, serious debility, or organ failure occurs. Slowly progressive diseases are also chronic diseases; many are also degenerative diseases. The opposite of progressive disease is *stable disease* or *static disease*: a medical condition that exists, but does not get better or worse.

Cure

A cure is the end of a medical condition or a treatment that is very likely to end it, while remission refers to the disappearance, possibly temporarily, of symptoms. Complete remission is the best possible outcome for incurable diseases.

Scope

Localized disease

A localized disease is one that affects only one part of the body, such as athlete's foot or an eye infection.

Disseminated disease

A disseminated disease has spread to other parts; with cancer, this is usually called metastatic disease.

Systemic disease

A systemic disease is a disease that affects the entire body, such as influenza or high blood pressure.

Causes and transmission

Only some diseases such as influenza are contagious and commonly believed infectious. The micro-organisms that cause these diseases are known as pathogens and include varieties of bacteria, viruses, protozoa and fungi. Infectious diseases can be transmitted, e.g. by hand-to-mouth contact with infectious material on surfaces, by bites of insects or other carriers of the disease, and from contaminated water or food (often via fecal contamination), etc In addition, there are sexually transmitted diseases. In some cases, microorganisms that are not readily spread from person to person play a role, while other diseases can be prevented or ameliorated with appropriate nutrition or other lifestyle changes.

Some diseases, such as most (but not all) forms of cancer, heart disease and mental disorders, are non-infectious diseases. Many non-infectious diseases have a partly or completely genetic basis and may thus be transmitted from one generation to another.

Social determinants of health are the social conditions in which people live that determine their health. Illnesses are generally related to social, economic, political, and environmental circumstances. Social determinants of health have been recognized by several health organizations such as the Public Health Agency of Canada and the World Health Organization to greatly influence collective and personal well-being. The World Health Organization's Social Determinants Council also recognizes Social determinants of health in poverty.

When the cause of a disease is poorly understood, societies tend to mythologize the disease or use it as a metaphor or symbol of whatever that culture considers evil. For example, until the bacterial cause of tuberculosis was discovered in 1882, experts variously ascribed the disease to heredity, a sedentary lifestyle, depressed mood, and overindulgence in sex, rich food, or alcohol—all the social ills of the time.

Disease burden

Disease burden is the impact of a health problem in an area measured by financial cost, mortality, morbidity, or other indicators.

There are several measures used to quantify the burden imposed by diseases on people. The years of potential life lost (YPLL) is a simple estimate of the number of years that a person's life was shortened due to a disease. For example, if a person dies at the age of 65 from a disease, and would probably have lived until age 80 without that disease, then that disease has caused a loss of 15 years of potential life. YPLL measurements do not account for how disabled a person is before dying, so the measurement treats a person who dies suddenly and a person who died at the same age after decades of illness as equivalent. In 2004, the World Health Organization calculated that 932 million years of potential life were lost to premature death.

The quality-adjusted life year (QALY) and disability-adjusted life year (DALY) metrics are similar, but take into account whether the person was healthy after diagnosis. In addition to the number of years lost due to premature death, these measurements add part of the years lost to being sick. Unlike YPLL, these measurements show the burden imposed on people who are very sick, but who live a normal lifespan. A disease that has high morbidity, but low mortality, has a high DALY and a low YPLL. In 2004, the World Health Organization calculated that 1.5 billion disability-adjusted life years were lost to disease and injury.

In the developed world, heart disease and stroke cause the most loss of life, but neuropsychiatric conditions like major depressive disorder cause the most years lost to being sick.

The most common diseases and their quality-adjusted life year (QALY) and disability-adjusted life year (DALY) are given below in the form of table.

Figure 1 Common diseases and their quality-adjusted life year (QALY) and disability-adjusted life year (DALY)

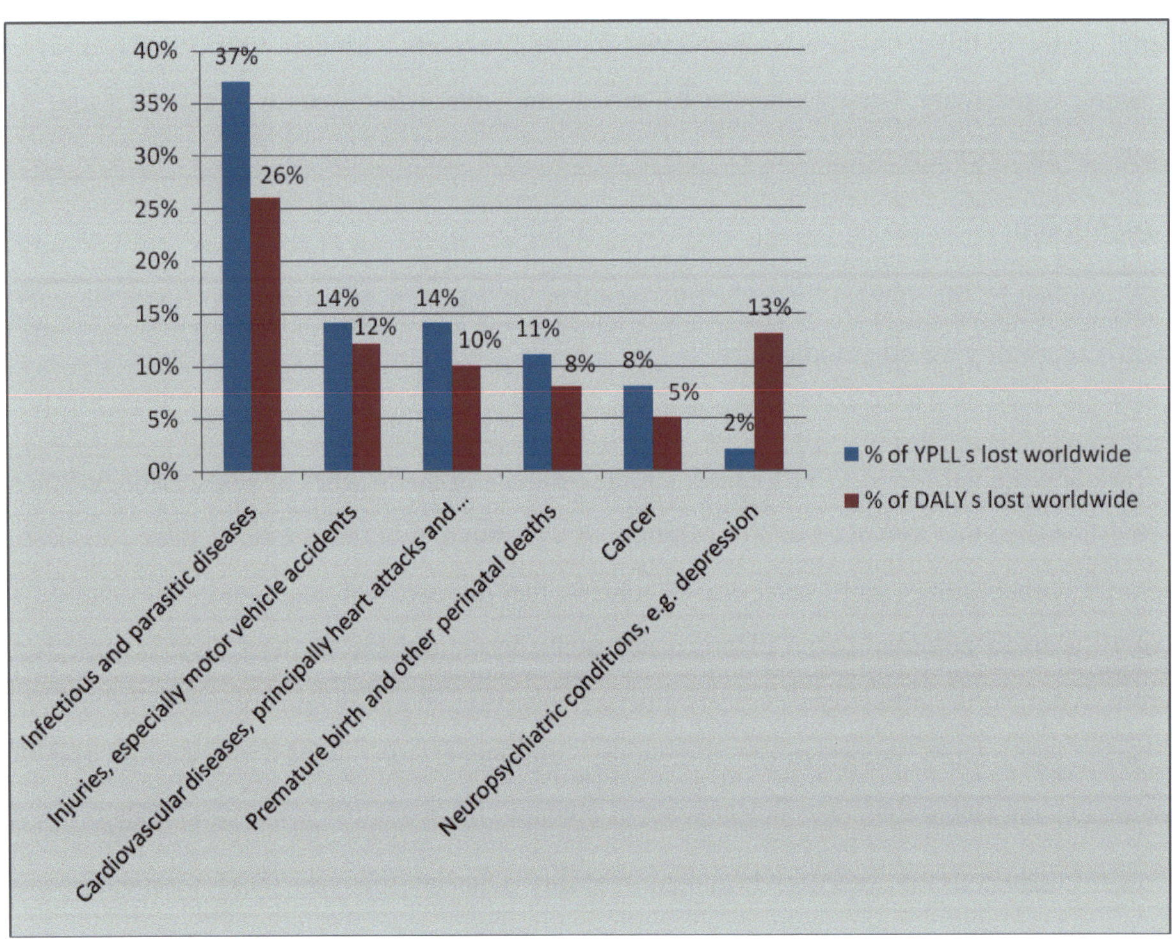

Therapy

Medical therapies or treatments are efforts to cure or improve a disease or other health problem. In the medical field, therapy is synonymous with the word treatment. Among psychologists, the term may refer specifically to psychotherapy or "talk therapy". Common treatments include medications, surgery, medical devices, and self-care. Treatments may be provided by an organized health care system, or informally, by the patient or family members.

A prevention or preventive therapy is a way to avoid an injury, sickness, or disease in the first place. A treatment or cure is applied after a medical problem has already started. A treatment attempts to improve or remove a problem, but treatments may not produce permanent cures, especially in chronic diseases. Cures are a subset of treatments that reverse diseases completely or end medical problems permanently. Many diseases that cannot be completely cured are still treatable. Pain management (also called pain medicine) is that branch of medicine employing an interdisciplinary approach to the relief of pain and improvement in the quality of life of those living with pain

Treatment for medical emergencies must be provided promptly, often through an emergency department or, in less critical situations, through an urgent care facility.

CHAPTER 2 LIST OF THE 5 DEADLY DISEASES

Sorting out the 5 deadliest diseases is a difficult task. In this volume the list which would be given are the top most deadly diseases present on this world. In coming volumes more deadly diseases would be given.

The list of 5 deadliest diseases are as follows :-

1. Ebola Virus
2. Marbung Virus
3. NIPAH virus
4. Rabies
5. H5N1 Virus

CHAPTER 3 EBOLA VIRUS

Ebola virus (EBOV) causes extremely severe disease in humans and in nonhuman primates in the form of viral hemorrhagic fever. EBOV is a select agent, World Health Organization Risk Group 4 Pathogen (requiring Biosafety Level 4-equivalent containment), National Institutes of Health/National Institute of Allergy and Infectious Diseases Category A Priority Pathogen, Centers for Disease Control and Prevention Category A Bioterrorism Agent, and listed as a Biological Agent for Export Control by the Australia Group.

Infection typically begins with flu-like symptoms, which often progress rapidly to fatal complications of hemorrhage, fever, and hypotensive shock.
The Ebola virus was first identified in the western equatorial province of Sudan and in a nearby region of Zaire (now Democratic Republic of the Congo) in 1976 after significant epidemics in Nzara, southern Sudan and Yambuku, northern Zaire.

There are five distinct species of the Ebola virus: Bundibugyo, Côte d'Ivoire, Reston, Sudan and Zaïre. Bundibugyo, Sudan and Zaïre species have been associated with large outbreaks of Ebola haemorrhagic fever (EHF) in Africa causing death in 25-90% of all clinically ill cases, while Côte d'Ivoire and Reston have not.

The Ebola virus is transmitted by direct contact with the blood, body fluids and tissues of infected persons. Transmission of the Ebola virus has also occurred by handling sick or dead infected wild animals (chimpanzees, gorillas, monkeys, forest antelope, fruit bats). The predominant treatment is general supportive therapy.

History of Ebola

Ebola virus (abbreviated EBOV) was first described in 1976 by David Finkes. In 1976, Ebola (named after the Ebola River in Zaire) first emerged in Sudan and Zaire. The first outbreak of Ebola (Ebola-Sudan) infected over 284 people, with a mortality rate of 53%. A few months later, the second Ebola virus emerged from Yambuku, Zaire, Ebola-Zaire (EBOZ). EBOZ, with the highest mortality rate of any of the Ebola viruses (88%), infected 318 people. Despite the tremendous effort of experienced and dedicated researchers, Ebola's natural reservoir was never identified. The third strain of Ebola, Ebola Reston (EBOR), was first identified in 1989 when infected monkeys were imported into Reston, Virginia, from Mindanao in the Philippines. Fortunately, the few people who were infected with EBOR (seroconverted) never developed Ebola hemorrhagic fever (EHF). The last known strain of Ebola, Ebola Cote d'Ivoire (EBO-CI) was discovered in 1994 when a female ethologist performing a necropsy on a dead chimpanzee from the Tai Forest, Cote d'Ivoire, accidentally infected herself during the necropsy.

Classification of Ebola

The genera Ebolavirus and Marburgvirus were originally classified as the species of the now-obsolete Filovirus genus. In March 1998, the Vertebrate Virus Subcommittee proposed in the International Committee on Taxonomy of Viruses (ICTV) to change the Filovirusgenus to the Filoviridae family with two specific genera: Ebola-like viruses and Marburg-like viruses. This proposal was implemented in Washington, DC on April 2001 and in Paris on July 2002. In 2000, another proposal was made in Washington, DC, to change the "-like viruses" to "-virus" resulting in today's Ebolavirus and Marburgvirus.

Ebola species subtypes

The five characterised Ebola species are:

Zaire ebolavirus (ZEBOV)

Also known simply as the *Zaire virus*, ZEBOV has the highest case-fatality rate of the ebolaviruses, up to 90% in some epidemics, with an average case fatality rate of approximately 83% over 27 years. There have been more outbreaks of *Zaire ebolavirus* than of any other species. The first outbreak occurred on 26 August 1976 in Yambuku. The first recorded case was Mabalo Lokela, a 44-year-old schoolteacher. The symptoms resembled malaria, and subsequent patients received quinine. Transmission has been attributed to reuse of unsterilized needles and close personal contact.

Sudan ebolavirus (SEBOV)

Like the *Zaire virus*, SEBOV emerged in 1976; it was at first assumed to be identical with the Zaire species. SEBOV is believed to have broken out first among cotton factory workers in Nzara, Sudan, with the first case reported as a worker exposed to a potential natural reservoir. The virus was not found in any of the local animals and insects that were tested in response. The carrier is still unknown. The lack of barrier nursing (or "bedside isolation") facilitated the spread of the disease. The most recent outbreak occurred in May, 2004. Twenty confirmed cases were reported in Yambio County, Sudan, with five deaths resulting. The average fatality rates for SEBOV were 54% in 1976, 68% in 1979, and 53% in 2000 and 2001.

Reston ebolavirus (REBOV)

Discovered during an outbreak of simian hemorrhagic fever virus (SHFV) in crab-eating macaques from Hazleton Laboratories (now Covance) in 1989. Since the initial outbreak in Reston, Virginia, it has since been found in non-human primates in Pennsylvania, Texas and Siena, Italy. In each case, the affected animals had been imported from a facility in the Philippines, where the virus has also infected pigs. Despite its status as a Level-4 organism and its apparent pathogenicity in monkeys, REBOV did not cause disease in exposed human laboratory workers.

Côte d'Ivoire ebolavirus (CIEBOV)

Also referred to as *Taï Forest ebolavirus* and by the English place name, "Ivory Coast", it was first discovered among chimpanzees from the Taï Forest in Côte d'Ivoire, Africa, in 1994. Necropsies showed blood within the heart to be brown; no obvious marks were seen on the organs; and one necropsy showed lungs filled with blood. Studies of tissues taken from the chimpanzees showed results similar to human cases during the 1976 Ebola outbreaks in Zaire and Sudan. As more dead chimpanzees were discovered, many tested positive for Ebola using molecular techniques. The source of the virus was believed to be the meat of infected Western Red Colobus monkeys, upon which the chimpanzees preyed. One of the scientists performing the necropsies on the infected chimpanzees contracted Ebola. She developed symptoms similar to those of dengue fever approximately a week after the necropsy, and was transported to Switzerland for treatment. She was discharged from the hospital after two weeks and had fully recovered six weeks after the infection.

Bundibugyo ebolavirus (BEBOV)

On 24 November 2007, the Uganda Ministry of Health confirmed an outbreak of Ebolavirus in the Bundibugyo District. After confirmation of samples tested by the United States National Reference Laboratories and the CDC, the World Health Organization confirmed the presence of the new species. On 20 February 2008, the Uganda Ministry officially announced the end of the epidemic in Bundibugyo, with the last infected person discharged on 8 January 2008.[11] An epidemiological study conducted by WHO and Uganda Ministry of Health scientists determined there were 116 confirmed and probable cases of the new Ebola species, and that the outbreak had a mortality rate of 34% (39 deaths). In 2012, there was an outbreak of Bundibugyo ebolavirus in a northeastern province of the Democratic Republic of the Congo. There were 15 confirmed cases and 10 fatalities.

Signs and symptoms of Ebola hemorrhagic fever

Ebola Virus Disease begins with a sudden onset of an influenza-like stage characterized by general malaise, fever with chills, arthralgia,myalgia, and chest pain. Nausea is accompanied by abdominal pain, diarrhea, and vomiting. Respiratory tract involvement is characterized by pharyngitis with sore throat, cough, dyspnea, and hiccups. The central nervous system is affected as judged by the development of severe headaches, agitation, confusion, fatigue, depression, seizures, and sometimes coma.

Cutaneous presentation may include: maculopapular rash, petechiae, purpura, ecchymoses, and hematomas (especially around needle injection sites). Development of hemorrhagic symptoms is generally indicative of a negative prognosis. However, contrary to popular belief, hemorrhage does not lead to hypovolemia and is not the cause of death (total blood loss is low except during labor). Instead, death occurs due to multiple organ dysfunction syndrome (MODS) due to fluid redistribution, hypotension, disseminated intravascular coagulation, and focal tissue necroses.

The mean incubation period, best calculated currently for EVD outbreaks due to EBOV infection, is 12.7 days (standard deviation = 4.3 days), but can be as long as 25 days

Risk factors

Between 1976 and 1998, from 30,000 mammals, birds, reptiles, amphibians, and arthropods sampled from outbreak regions, no *ebola virus* was detected apart from some genetic traces found in six rodents (*Mus setulosus* and *Praomys*) and one shrew (*Sylvisorex ollula*) collected from the Central African Republic. Traces of EBOV were detected in the carcasses of gorillas and chimpanzees during outbreaks in 2001 and 2003, which later became the source of human infections. However, the high lethality from infection in these species makes them unlikely as a natural reservoir.

Plants, arthropods, and birds have also been considered as possible reservoirs; however, bats are considered the most likely candidate. Bats were known to reside in the cotton factory in which the index cases for the 1976 and 1979 outbreaks were employed, and they have also been implicated in Marburg virus infections in 1975 and 1980. Of 24 plant species and 19 vertebrate species experimentally inoculated with EBOV, only bats became infected. The absence of clinical signs in these bats is characteristic of a reservoir species. In a 2002–2003 survey of 1,030 animals which included 679 bats from Gabon and the Republic of the Congo, 13 fruit bats were found to contain EBOV RNA fragments.[1] As of 2005, three types of fruit bats (*Hypsignathus monstrosus*, *Epomops franqueti*, and *Myonycteris torquata*) have been identified as being in contact with EBOV. They are now suspected to represent the EBOV reservoir hosts.

The existence of integrated genes of filoviruses in some genomes of small rodents, insectivorous bats, shrews, tenrecs, and marsupials indicates a history of infection with filoviruses in these groups as well. However, it has to be stressed that infectious ebolaviruses have not yet been isolated from any nonhuman animal.

Bats drop partially eaten fruits and pulp, then terrestrial mammals such as gorillas and duikers feed on these fallen fruits. This chain of events forms a possible indirect means of transmission from the natural host to animal populations, which have led to research towards viral shedding in the saliva of bats. Fruit production, animal behavior, and other factors vary at different times and places which may trigger outbreaks among animal populations. Transmission between natural reservoirs and humans are rare, and outbreaks are usually traceable to a single index case where an individual has handled the carcass of gorilla, chimpanzee, or duiker The virus then spreads person-to-person, especially within families, hospitals, and during some mortuary rituals where contact among individuals becomes more likely

The virus has been confirmed to be transmitted through body fluids. Transmission through oral exposure and through conjunctiva exposure is likely and has been confirmed in non-human primates. Filoviruses are not naturally transmitted by aerosol. They are, however, highly infectious as breathable 0.8–1.2 micrometre droplets in laboratory conditions; because of this potential route of infection, these viruses have been classified as Category A biological weapons.

All epidemics of Ebola have occurred in sub-optimal hospital conditions, where practices of basic hygiene and sanitation are often either luxuries or unknown to caretakers and where disposable needles and autoclaves are unavailable or too expensive. In modern hospitals with disposable needles and knowledge of basic hygiene and barrier nursing techniques, Ebola has never spread on a large scale. In isolated settings such as a quarantined hospital or a remote village, most victims are infected shortly after the first case of infection is present. The quick onset of symptoms from the time the disease becomes contagious in an individual makes it easy to identify sick individuals and limits an individual's ability to spread the disease by traveling. Because bodies of the deceased are still infectious, some doctors had to take measures to properly dispose of dead bodies in a safe manner despite local traditional burial rituals

Virology

Genome

Like all mononegaviruses, ebolavirions contain linear nonsegmented, single-stranded, non-infectious RNA genomes of negative polarity that possesses inverse-complementary 3' and 5' termini, do not possess a 5' cap, are not polyadenylated, and are not covalently linked to a protein Ebolavirus genomes are approximately 19 kilobase pairs long and contain seven genes in the order 3'-UTR-NP-VP35-VP40-GP-VP30-VP24-L-5'-UTR. The genomes of the five different ebolaviruses (BDBV, EBOV, RESTV, SUDV, and TAFV) differ in sequence and the number and location of gene overlaps.

Structure

Like all filoviruses, ebolavirions are filamentous particles that may appear in the shape of a shepherd's crook or in the shape of a "U" or a "6", and they may be coiled, toroid, or branched Ebolavirions are generally 80 nm in width, but vary somewhat in length. In general, the median particle length of ebolaviruses ranges from 974–1,086 nm (in contrast to marburgvirions, whose median particle length was measured to be 795–828 nm), but particles as long as 14,000 nm have been detected in tissue culture. Ebolavirions consist of seven structural proteins. At the center is the helical ribonucleocapsid, which consists of the genomic RNA wrapped around a polymer of nucleoproteins (NP). Associated with the ribonucleoprotein is the RNA-dependent RNA polymerase (L) with the polymerase cofactor (VP35) and a transcription activator (VP30). The ribonucleoprotein is embedded in a matrix, formed by the major (VP40) and minor (VP24) matrix proteins. These particles are surrounded by a lipid membrane derived from the host cell membrane. The membrane anchors a glycoprotein ($GP_{1,2}$) that projects 7 to 10 nm spikes away from its surface. While nearly identical to marburgvirions in structure, ebolavirions are antigenically distinct.

Entry

Niemann–Pick C1 (NPC1) appears to be essential for Ebola infection. Two independent studies reported in the same issue of Nature showed that Ebola virus cell entry and replication requires the cholesterol transporter protein NPC1. When cells from Niemann Pick Type C1 patients (who have a mutated form of NPC1) were exposed to Ebola virus in the laboratory, the cells survived and appeared immune to the virus, further indicating that Ebola relies on NPC1 to enter cells. This might imply that genetic mutations in the NPC1 gene in humans could make some people resistant to one of the deadliest known viruses affecting humans. The same studies described similar results with Ebola's cousin in the filovirus group, Marburg virus, showing that it too needs NPC1 to enter cells. Furthemore, NPC1 was shown to be critical to filovirus entry because it mediates infection by binding directly to the viral envelope glycoprotein. A later study confirmed the findings that NPC1 is a critical filovirus receptor that mediates infection by binding directly to the viral envelope glycoprotein and that the second lysosomal domain of NPC1 mediates this binding

In one of the original studies, a small molecule was shown to inhibit Ebola virus infection by preventing the virus glycoprotein from binding to NPC1. In the other study, mice that were heterozygous for NPC1 were shown to be protected from lethal challenge with mouse adapted Ebola virus . Together, these studies suggest NPC1 may be potential therapeutic target for an Ebola anti-viral drug.

Replication

The ebolavirus life cycle begins with virion attachment to specific cell-surface receptors, followed by fusion of the virion envelope with cellular membranes and the concomitant release of the virus nucleocapsid into the cytosol. The viral RNA polymerase, encoded by the L gene, partially uncoats the nucleocapsid and transcribes the genes into positive-stranded mRNAs, which are then translated into structural and nonstructural proteins. Ebolavirus RNA polymerase (L) binds to a single promoter located at the 3' end of the genome. Transcription either terminates after a gene or continues to the next gene downstream. This means that genes close to the 3' end of the genome are transcribed in the greatest abundance, whereas those toward the 5' end are least likely to be transcribed. The gene order is therefore a simple but effective form of transcriptional regulation. The most abundant protein produced is the nucleoprotein, whose concentration in the cell determines when L switches from gene transcription to genome replication. Replication results in full-length, positive-stranded antigenomes that are, in turn, transcribed into negative-stranded virus progeny genome copy. Newly synthesized structural proteins and genomes self-assemble and accumulate near the inside of the cell membrane. Virions bud off from the cell, gaining their envelopes from the cellular membrane they bud from. The mature progeny particles then infect other cells to repeat the cycle

Pathophysiology

Endothelial cells, mononuclear phagocytes, and hepatocytes are the main targets of infection. After infection, in a secreted glycoprotein (sGP) the Ebola virus glycoprotein (GP) is synthesized. Ebola replication overwhelms protein synthesis of infected cells and host immune defenses. The GP forms a trimeric complex, which binds the virus to the endothelial cells lining the interior surface of blood vessels. The sGP forms a dimeric protein which interferes with the signaling of neutrophils, a type of white blood cell, which allows the virus to evade the immune system by inhibiting early steps of neutrophil activation. These white blood cells also serve as carriers to transport the virus throughout the entire body to places such as the lymph nodes, liver, lungs, and spleen. The presence of viral particles and cell damage resulting from budding causes the release of cytokines (specifically TNF-α, IL-6, IL-8, etc.), which are the signaling molecules for fever and inflammation. The cytopathic effect, from infection in the endothelial cells, results in a loss of vascular integrity. This loss in vascular integrity is furthered with synthesis of GP, which reduces specific integrins responsible for cell adhesion to the inter-cellular structure, and damage to the liver, which leads to coagulopathy

Diagnosis

VD is clinically indistinguishable from Marburg virus disease (MVD), and it can also easily be confused with many other diseases prevalent in Equatorial Africa, such as other viral hemorrhagic fevers, falciparum malaria, typhoid fever, shigellosis, rickettsial diseases such as typhus, cholera, gram-negative septicemia, borreliosis such as relapsing fever or EHEC enteritis. Other infectious diseases that ought to be included in the differential diagnosis include leptospirosis, scrubtyphus, plague, candidiasis, histoplasmosis, trypanosomiasis, visceral leishmaniasis, hemorrhagic smallpox, measles, and fulminant viral hepatitis. Non-infectious diseases that can be confused with EVD are acute promyelocytic leukemia, hemolytic uremic syndrome, snake envenomation, clotting factor deficiencies/platelet disorders, thrombotic thrombocytopenic purpura, hereditary hemorrhagic telangiectasia, Kawasaki disease, and even warfarin intoxication.

Prevention

Ebola viruses are highly infectious as well as contagious.

As an outbreak of ebola progresses, bodily fluids from diarrhea, vomiting, and bleeding represent a hazard. Due to lack of proper equipment and hygienic practices, large-scale epidemics occur mostly in poor, isolated areas without modern hospitals or well-educated medical staff. Many areas where the infectious reservoir exists have just these characteristics. In such environments, all that can be done is to immediately cease all needle-sharing or use without adequate sterilization procedures, isolate patients, and observe strict barrier nursing procedures with the use of a medical-rated disposable face mask, gloves, goggles, and a gown at all times, strictly enforced for all medical personnel and visitors. The aim of all of these techniques is to avoid any person's contact with the blood or secretions of any patient, including those who are deceased.

Vaccines have successfully protected nonhuman primates; however, the six months needed to complete immunization made it impractical in an epidemic. To resolve this, in 2003, a vaccine using an adenoviral (ADV) vector carrying the Ebola spike protein was tested on crab-eating macaques. The monkeys were challenged with the virus 28 days later, and remained resistant. In 2005, a vaccine based on attenuated recombinant vesicular stomatitis virus (VSV) vector carrying either the Ebola glycoprotein or Marburg glycoprotein successfully protected nonhuman primates, opening clinical trials in humans. By October, the study completed the first human trial; giving three vaccinations over three months showing capability of safely inducing an immune response. Individuals were followed for a year, and, in 2006, a study testing a faster-acting, single-shot vaccine began. This study was completed in 2008. The next step is to try the vaccine on a strain of Ebola that is closer to the one that infects humans. There are currently no Food and Drug Administration-approved vaccines for the prevention of EVD. Many candidate vaccines have been developed and tested in various animal models. Of those, the most promising ones are DNA vaccines or are based on adenoviruses, vesicular stomatitis Indiana virus (VSIV) or filovirus-like particles (VLPs) as all of these candidates could protect nonhuman primates from ebolavirus-induced disease. DNA vaccines, adenovirus-based vaccines, and VSIV-based vaccines have entered clinical trials.

Contrary to popular belief, ebolaviruses are not transmitted by aerosol during natural EVD outbreaks. Due to the absence of an approved vaccine, prevention of EVD therefore relies predominantly on behavior modification, proper personal protective equipment, and sterilization/disinfection.

On 6 December 2011, the development of a successful vaccine against Ebola for mice was reported. Unlike the predecessors, it can be freeze-dried and thus stored for long periods in wait for an outbreak. The research is reported in *Proceedings of National Academy of Sciences*

According to the authors (Phoolcharoen et al.) of the above article they have evaluated the immunogenicity an efficacy of an EBOV vaccine candidate in which the viral surface glycoprotein is biomanufactured as a fusion to a monoclonal antibody that recognizes an epitope in glycoprotein, resulting in the production of Ebola immune complexes (EICs).

Although antigen– antibody immune complexes are known to be efficiently processed and presented to immune effector cells, authors found that codelivery of the EIC with Toll-like receptor agonists elicited a more robust antibody response in mice than did EIC alone. Among the compounds tested by authors , polyinosinic:polycytidylic acid (PIC, a Toll-like receptor 3 agonist) was highly effective as an adjuvant agent.

After vaccinating mice with EIC plus PIC, 80% of the animals were protected against a lethal challenge with live EBOV (30,000 LD50 of mouse adapted virus). Surviving animals showed a mixed Th1/Th2 response to the antigen, suggesting this may be important for protection. Survival after vaccination with EIC plus PIC was statistically equivalent to that achieved with an alternative viral vector vaccine candidate reported in the literature. Because nonreplicating subunit vaccines offer the possibility of formulation for cost effective, long-term storage in bio threat reduction repositories, EIC is an attractive option for public health defense measures.

Recent research

In late 2012, Canadian scientists discovered that the deadliest form of the virus could be transmitted by air between species. They managed to prove that the virus was transmitted from pigs to monkeys without any direct contact between them, leading to fears that airborne transmission could be contributing to the wider spread of the disease in parts of Africa. Evidence was also found that pigs might be one of the reservoir hosts for the virus; the fruit bat has long been considered as the reservoir

CHAPTER 4 MARBURG VIRUS

Introduction

Marburg virus is a hemorrhagic fever virus first noticed and described during small epidemics in the German cities Marburg and Frankfurt and the Yugoslavian capital Belgrade in the 1960s. Workers were accidentally exposed to tissues of infected grivets (*Chlorocebus aethiops*) at the city's former main industrial plant, the Behringwerke, then part of Hoechst, and today of CSL Behring. During these outbreaks, 31 people became infected and seven of them died. Marburg virus (MARV) causes severe disease in humans and nonhuman primates in the form of viral hemorrhagic fever. MARV is a Select Agent, WHO Risk Group 4 Pathogen (requiring biosafety level 4-equivalent containment), NIH/National Institute of Allergy and Infectious Diseases Category A Priority Pathogen, Centers for Disease Control and Prevention Category A Bioterrorism Agent and is listed as a biological agent for export control by the Australia Group.

Marburgvirus (the gentle sister) affects humans somewhat like nuclear radiation, damaging virtually all of the tissues in their bodies. It attacks with particular ferocity the internal organs, connective tissue, intestines, and skin. In Germany, all the survivors lost their hair-they went bald or partly bald. Their hair died at the roots and fellout in clumps, as if they had received radiation burns. Hemorrhage occurred from all orifices of the body. During the survivors' recovery period, the skin peeled off their faces, hands, feet, and genitals. Some of the men suffered from blown up, semirotten testicles. One of the worst cases of this appeared in a morgue attendant who had handled Marburg-infected bodies. The virus also lingered in the fluid inside the eyeballs of some victims for many months

History

The year was 1967. Several laboratory workers, all from the same lab in Marburg, Germany, were hospitalized with a severe and strange disease. The physicians on staff realized the workers were all suffering from the same ailment, with symptoms that included fever, diarrhea, vomiting, massive bleeding from many different organs, shock, and eventually circulatory system collapse. An investigation began in an attempt to uncover the source of the outbreak. This led to the identification of the source of the virus in Germany: a species of African green monkeys, imported from Uganda, which were being used by the scientists for polio vaccine research. The virus was isolated, and found to exhibit a unique morphology, leading to the designation of a new group: the Filoviridae In that outbreak, a total of 31 human cases were observed, and the disease presented with a 23% mortality rate (7 deaths occurred out of 31 total infections).

After this episode, the virus went back into hiding for almost a decade, not surfacing again until 1975 in South Africa. The origin of this outbreak is unknown, although based on epidemiological studies it is assumed that the index case (the first person known to have been infected), a young man hitchhiking through Africa, acquired the disease in Zimbabwe, and then infected 2 others in South Africa when he arrived there. Only the index case died from the disease; the secondary cases (those infected due to contact with the index case) survived.

Marburg again disappeared until one case was reported in Kenya in 1980, and again in 1987 in the same area. In the first outbreak, again only the index case died, while a second patient survived. Only one infection was noted in the 1987 outbreak, resulting in death of the patient. Both of these Kenyan outbreaks occurred in the vicinity of a volcano named Mount Elgon, and there is evidence both index cases had spent time in a cave inside the mountain. This has led to (unconfirmed) speculation that bats may be a reservoir for filoviruses;

Between 1987 and 1998, the only cases of Marburg were due to laboratory accidents, both in the former Soviet Union. One of these cases was fatal. However, in 1998, the largest natural outbreak of Marburg virus disease to date began in northeastern Democratic Republic of the Congo (DRC). This time, the focus of the outbreak was a town called Durba (population 16,000). A large number of men in this region work for the Kilo Moto Mining Company, which runs a number of illegal gold mines in the area. Working in this area is precarious; civil war broke out in 1996, and the socio-economic situation has deteriorated since then. Infectious diseases of all types are common, as vaccinations and medication are in short supply. The Marburg outbreak is thought to have started in November of 1998, although it was not reported to any international agencies until late April of 1999, following the death of the chief medical officer in the area from the disease. At this time, local officials contacted the Medecins sans Frontieres (Doctors without Borders) in Belgium regarding the ongoing hemorrhagic fever epidemic. Both this group and their sister group in Holland sent officers to investigate and to stem the spread of the epidemic.

Patient samples were immediately sent to the National Institute of Virology in Johannesburg, South Africa; a diagnosis of Marburg virus as the cause of the illness was made on May 6th. Barrier nursing procedures were instituted, and isolation wards were fashioned at the hospital. Over the course of the epidemic, 149 cases with 123 deaths were recorded (83% fatality rate). Miners were found to be at a significantly higher risk of contracting Marburg than the general population of this area, suggesting they may be more frequently exposed to the natural reservoir of Marburg virus.

This outbreak was eclipsed beginning in 2004, when Marburg hit Angola in the fall. Due to civil war in the country and a non-existent public health infrastructure, the outbreak wasn't identified until almost 6 months later, in March of 2005. 252 people were infected, with 227 deaths (90% fatality rate). The outbreak was declared over in November of 2005, after no additional cases had been reported since summer.

This brings us to present-day Uganda, where another Marburg outbreak is ongoing, again associated with miners. So far, this one has been minor, and seems to be contained; meanwhile, back in the DRC, there comes a report of another outbreak of hemorrhagic fever. It's not known at the time what the causative agent is, but a hundred deaths have already been reported:

Classification

Marburg virus disease (MVD) is the official name listed in the World Health Organization's International Statistical Classification of Diseases and Related Health Problems 10 (ICD-10) for the human disease caused by any of the two marburgviruses Marburg virus(MARV) and Ravn virus (RAVV). In the scientific literature, Marburg hemorrhagic fever (MHF) is often used as an unofficial alternative name for the same disease. Both disease names are derived from the German city Marburg, where MARV was first discovered

Causes

MVD is caused by two viruses classified in the genus Marburgvirus, family Filoviridae, order Mononegavirales: Marburg virus (MARV) and Ravn virus (RAVV).

Marburgviruses are endemic in arid woodlands of equatorial Africa. Most marburgvirus infections were repeatedly associated with people visiting natural caves or working in mines. In 2009, the successful isolation of infectious MARV and RAVV was reported from healthy Egyptian rousettes (Rousettus aegyptiacus) caught in caves.

This isolation strongly suggests that Old World fruit bats are involved in the natural maintenance of marburgviruses and that visiting bat-infested caves is a risk factor for acquiring marburgvirus infections. Further studies are necessary to establish whether Egyptian rousettes are the actual hosts of MARV and RAVV or whether they get infected via contact with another animal and therefore serve only as intermediate hosts. Another risk factor is contact with nonhuman primates, although only one outbreak of MVD (in 1967) was due to contact with infected monkeys.[1] Finally, a major risk factor for acquiring marburgvirus infection is occupational exposure, i.e. treating patients with MVD without proper personal protective equipment.

Contrary to Ebola virus disease (EVD), which has been associated with heavy rains after long periods of dry weather, triggering factors for spillover of marburgviruses into the human population have not yet been described.

Epidemiology

Epidemiology of Marburg virus infections

Year	Virus	Geographic location	Human cases/deaths (case-fatality rate)
1967	MARV	Marburg and Frankfurt, West Germany; and Belgrade, Yugoslavia	31/7 (23%)
1975	MARV	Rhodesia and Johannesburg, South Africa	3/1 (33%)

1980	MARV	Kenya	2/1 (50%)
1987	RAVV	Kenya	1/1 (100%)[
1988	MARV	Koltsovo, Soviet Union	1/1 (100%) [laboratory accident]
1990	MARV	Koltsovo, Soviet Union	1/0 (0%) [laboratory accident]
1998–2000	MARV + RAVV	Durba and Watsa, Democratic Republic of the Congo	154/128 (83%
2004–2005	MARV	Angola	252/227 (90%)
2007	MARV + RAVV	Uganda	4/1 (25%
2008	MARV	Uganda, Netherlands, Colorado	2/1 (50%)

MVD was first documented in 1967, when 31 people became ill in the German towns of Marburg and Frankfurt am Main, and in Belgrade, Yugoslavia. The outbreak involved 25 primary MARV infections and seven deaths, and six nonlethal secondary cases. The outbreak was traced to infected grivets (species *Chlorocebus aethiops*) imported from an undisclosed location in Uganda and used in developing poliomyelitis vaccines. The monkeys were received by Behringwerke, a Marburg company founded by the first winner of the Nobel Prize in Medicine, Emil von Behring. The company, which at the time was owned by Hoechst, was originally set up to develop sera against tetanus and diphtheria. Primary infections occurred in Behringwerke laboratory staff while working with grivet tissues or tissue cultures without adequate personal protective equipment. Secondary cases involved two physicians, a nurse, a post-mortem attendant, and the wife of a veterinarian. All secondary cases had direct contact, usually involving blood, with a primary case. Both physicians became infected through accidental skin pricks when drawing blood from patients.

Virology

Genome

Like all mononegaviruses, marburgvirions contain non-infectious, linear nonsegmented, single-stranded RNA genomes of negative polarity that possesses inverse-complementary 3' and 5' termini, do not possess a 5' cap, are not polyadenylated, and are not covalently linked to a protein. Marburgvirus genomes are approximately 19 kb long and contain seven genes in the order 3'-UTR-*NP*-*VP35*-*VP40*-*GP*-*VP30*-*VP24*-*L*-5'-UTR The genomes of the two different marburgviruses (MARV and RAVV) differ in sequence.

Structure

Like all filoviruses, marburgvirions are filamentous particles that may appear in the shape of a shepherd's crook or in the shape of a "U" or a "6", and they may be coiled, toroid, or branched. Marburgvirions are generally 80 nm in width, but vary somewhat in length. In general, the median particle length of marburgviruses ranges from 795–828 nm (in contrast to ebolavirions, whose median particle length was measured to be 974–1,086 nm), but particles as long as 14,000 nm have been detected in tissue culture. Marburgvirions consist of seven structural proteins. At the center is the helical ribonucleocapsid, which consists of the genomic RNA wrapped around a polymer of nucleoproteins (NP). Associated with the ribonucleoprotein is the RNA-dependent RNA polymerase (L) with the polymerase cofactor (VP35) and a transcription activator (VP30). The ribonucleoprotein is embedded in a matrix, formed by the major (VP40) and minor (VP24) matrix proteins. These particles are surrounded by a lipid membrane derived from the host cell membrane. The membrane anchors a glycoprotein ($GP_{1,2}$) that projects 7 to 10 nm spikes away from its surface. While nearly identical to ebolavirions in structure, marburgvirions are antigenically distinct.

Replication

The marburgvirus life cycle begins with virion attachment to specific cell-surface receptors, followed by fusion of the virion envelope with cellular membranes and the concomitant release of the virus nucleocapsid into the cytosol. The virus RdRp partially uncoats the nucleocapsid and transcribes the genes into positive-stranded mRNAs, which are then translated into structural and nonstructural proteins. Marburgvirus L binds to a single promoter located at the 3' end of the genome. Transcription either terminates after a gene or continues to the next gene downstream. This means that genes close to the 3' end of the genome are transcribed in the greatest abundance, whereas those toward the 5' end are least likely to be transcribed. The gene order is therefore a simple but effective form of transcriptional regulation. The most abundant protein produced is the nucleoprotein, whose concentration in the cell determines when L switches from gene transcription to genome replication. Replication results in full-length, positive-stranded antigenomes that are in turn transcribed into negative-stranded virus progeny genome copies. Newly synthesized structural proteins and genomes self-assemble and accumulate near the inside of the cell membrane. Virions bud off from the cell, gaining their envelopes from the cellular membrane they bud from. The mature progeny particles then infect other cells to repeat the cycle

There are currently no Food and Drug Administration-approved vaccines for the prevention of MVD. Many candidate vaccines have been developed and tested in various animal models Of those, the most promising ones are DNA vaccines or based on Venezuelan equine encephalitis virus replicons, vesicular stomatitis Indiana virus (VSIV) or filovirus-like particles (VLPs) as all of these candidates could protect nonhuman primates from marburgvirus-induced disease. DNA vaccines have entered clinical trials. Marburgviruses are highly infectious, but not very contagious. Importantly, and contrary to popular belief, marburgviruses do not get transmitted by aerosol during natural MVD outbreaks. Due to the absence of an approved vaccine, prevention of MVD therefore relies predominantly on behavior modification, proper personal protective equipment, and sterilization/disinfection.

The natural maintenance hosts of marburg viruses remain to be identified unequivocally. However, the isolation of both MARV and RAVV from bats and the association of several MVD outbreaks with bat-infested mines or caves strongly suggests that bats are involved in marburg virus transmission to humans. Avoidance of contact with bats and abstaining from visits to caves is highly recommended, but may not be possible for those working in mines or people dependent on bats as a food source.

Since marburgviruses are not spreading via aerosol, the most straightforward prevention method during MVD outbreaks is to avoid direct (skin-to-skin) contact with patients, their excretions and body fluids, or possibly contaminated materials and utensils. Patients ought to be isolated but still have the right to be visited by family members. Medical staff should be trained and apply strict barrier nursing techniques (disposable face mask, gloves, goggles, and a gown at all times). Traditional burial rituals, especially those requiring embalming of bodies, ought to be discouraged or modified, ideally with the help of local traditional healers.

Marburgviruses are World Health Organization Risk Group 4 Pathogens, requiring Biosafety Level 4-equivalent containment. Laboratory researchers have to be properly trained in BSL-4 practices and wear proper personal protective equipment.

Treatment

There is currently no effective marburgvirus-specific therapy for MVD. Treatment is primarily supportive in nature and includes minimizing invasive procedures, balancing fluids and electrolytes to counter dehydration, administration of anticoagulants early in infection to prevent or control disseminated intravascular coagulation, administration of procoagulants late in infection to control hemorrhaging, maintaining oxygen levels, pain management, and administration of antibiotics or antimycotics to treat secondary infections. Experimentally, recombinant vesicular stomatitis Indiana virus (VSIV) expressing the glycoprotein of MARV has been used successfully in nonhuman primate models as post-exposure prophylaxis. Novel, very promising, experimental therapeutic regimens rely on antisense technology: phosphorodiamidate morpholino oligomers (PMOs) targeting the MARV genome could prevent disease in nonhuman primates

CHAPTER 5 NIPAH

Introduction

Henipavirus is a genus of the family Paramyxoviridae, order Mononegavirales containing three established species: Hendra virus, Nipah virus and Cedar Virus. The henipaviruses are naturally harboured by Pteropid fruit bats (flying foxes) and some microbat species. Henipavirus is characterised by a largegenome, a wide host range, and their recent emergence as zoonotic pathogens capable of causing illness and death in domestic animals and humans.

In 2009, RNA sequences of three novel viruses in phylogenetic relationship to known Henipaviruses were detected in Eidolon helvum (the African Straw-coloured fruit bat) in Ghana. The finding of these novel putative Henipaviruses outside Australia and Asia indicates that the region of potential endemicity of Henipaviruses extends to Africa

NIPAH virus

Nipah virus was identified in April 1999, when it caused an outbreak of neurological and respiratory disease on pig farms in peninsular Malaysia, resulting in 257 human cases, including 105 human deaths and the culling of one million pigs In Singapore, 11 cases, including one death, occurred in abattoir workers exposed to pigs imported from the affected Malaysian farms. The Nipah virus has been classified by the Centers for Disease Control and Prevention as a Category C agent.[104] The name "Nipah" refers to the place, Kampung Baru Sungai Nipah in Negeri Sembilan State, Malaysia, the source of the human case from which Nipah virus was first isolated.

The outbreak was originally mistaken for Japanese encephalitis (JE), however, physicians in the area noted that persons who had been vaccinated against JE were not protected, and the number of cases among adults was unusual Despite the fact that these observations were recorded in the first month of the outbreak, the Ministry of Health failed to react accordingly, and instead launched a nationwide campaign to educate people on the dangers of JE and its vector, Culex mosquitoes.

Symptoms of infection from the Malaysian outbreak were primarily encephalitic in humans and respiratory in pigs. Later outbreaks have caused respiratory illness in humans, increasing the likelihood of human-to-human transmission and indicating the existence of more dangerous strains of the virus.

Based on seroprevalence data and virus isolations, the primary reservoir for Nipah virus was identified as Pteropid fruit bats, including *Pteropus vampyrus* (Large Flying Fox), and *Pteropus hypomelanus* (Small Flying-fox), both of which occur in Malaysia.

The transmission of Nipah virus from flying foxes to pigs is thought to be due to an increasing overlap between bat habitats and piggeries in peninsular Malaysia. At the index farm, fruit orchards were in close proximity to the piggery, allowing the spillage of urine, faeces and partially eaten fruit onto the pigs. Retrospective studies demonstrate that viral spillover into pigs may have been occurring in Malaysia since 1996 without detection. During 1998, viral spread was aided by the transfer of infected pigs to other farms, where new outbreaks occurred.

Figure 2 Pteropus vampyrus (Large flying fox)

Outbreaks

Figure 3 Outbreak of NIPAH VIRUS

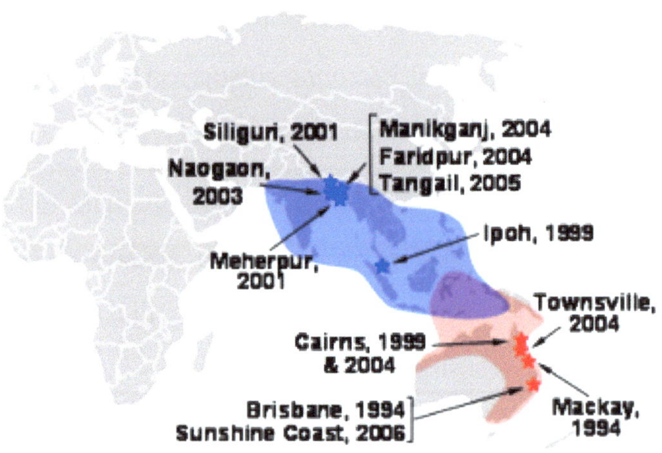

Eight more outbreaks of Nipah virus have occurred since 1998, all within Bangladesh and neighbouring parts of India. The outbreak sites lie within the range of Pteropus species (*Pteropus giganteus*). As with Hendra virus, the timing of the outbreaks indicates a seasonal effect. Cases occurring in Bangladesh during the winters of 2001, 2003, and 2004, were determined to have been caused by the Nipah virus. In February 2011, a Nipah outbreak began at Hatibandha Upazila in the Lalmonirhat Districtof northern Bangladesh. To date (7 February 2011), there have been 24 cases and 17 deaths in this outbreak.

- 2001 January 31–23 February, Siliguri, India: 66 cases with a 74% mortality rate. 75% of patients were either hospital staff or had visited one of the other patients in hospital, indicating person-to-person transmission.
- 2001 April – May, Meherpur District, Bangladesh: 13 cases with nine fatalities (69% mortality).
- 2003 January, Naogaon District, Bangladesh: 12 cases with eight fatalities (67% mortality).
- 2004 January – February, Manikganj and Rajbari districts, Bangladesh: 42 cases with 14 fatalities (33% mortality).
- 2004 19 February – 16 April, Faridpur District, Bangladesh: 36 cases with 27 fatalities (75% mortality). 92% of cases involved close contact with at least one other

person infected with Nipah virus. Two cases involved a single short exposure to an ill patient, including a rickshaw driver who transported a patient to hospital. In addition, at least six cases involved acute respiratory distress syndrome which has not been reported previously for Nipah virus illness in humans. This symptom is likely to have assisted human-to-human transmission through large droplet dispersal

- 2005 January, Tangail District, Bangladesh: 12 cases with 11 fatalities (92% mortality). The virus was probably contracted from drinking date palm juice contaminated by fruit bat droppings or saliva.[112]
- 2007 February – May, Nadia District, India: up to 50 suspected cases with 3–5 fatalities. The outbreak site borders the Bangladesh district of Kushtia where eight cases of Nipah virus encephalitis with five fatalities occurred during March and April 2007. This was preceded by an outbreak in Thakurgaon during January and February affecting seven people with three deaths. All three outbreaks showed evidence of person-to-person transmission.
- 2008 February – March, Manikganj and Rajbari districts, Bangladesh: Nine cases with eight fatalities.[114]
- 2010 January, Bhanga subdistrict, Faridpur, Bangladesh: Eight cases with seven fatalities. During March, one physician of Faridpur Medical College Hospital caring for confirmed Nipah cases died[115]
- 2011 February: An outbreak of Nipah Virus has occurred at Hatibandha, Lalmonirhat, Bangladesh. The deaths of 21 schoolchildren due to Nipah virus infection were recorded on 4 February 2011. IEDCR has confirmed the infection is due to this virus. Local schools were closed for one week to prevent the spread of the virus. People were also requested to avoid consumption of uncooked fruits and fruit products. Such foods, contaminated with urine or saliva from infected fruit bats, were the most likely source of this outbreak.

Nipah virus has been isolated from Lyle's flying fox (*Pteropus lylei*) in Cambodia and viral RNA found in urine and saliva from *P. lylei* and Horsfield's roundleaf bat (*Hipposideros larvatus*) inThailand.

Infective virus has also been isolated from environmental samples of bat urine and partially eaten fruit in Malaysia. Antibodies to henipaviruses have also been found in fruit bats in Madagascar (*Pteropusrufus, Eidolon dupreanum*) and Ghana (*Eidolon helvum*) indicating a wide geographic distribution of the viruses. No infection of humans or other species have been observed in Cambodia, Thailand or Africa thus far.

Pathology

In humans, the infection presents as fever, headache and drowsiness. Cough, abdominal pain, nausea, vomiting, weakness, problems with swallowing and blurred vision are relatively common. About a quarter of the patients have seizures and about 60% become comatose and might need mechanical ventilation. In patients with severe disease, their conscious state may deteriorate and they may develop severe hypertension, fast heart rate, and very high temperature.

Nipah virus is also known to cause relapse encephalitis. In the initial Malaysian outbreak, a patient presented with relapse encephalitis some 53 months after his initial infection. There is no definitive treatment for Nipah encephalitis, apart from supportive measures, such as mechanical ventilation and prevention of secondary infection. Ribavirin, an antiviral drug, was tested in the Malaysian outbreak, and the results were encouraging, though further studies are still needed.

While no vaccine currently exists, a recent (2012) study of a trial vaccine developed using the outer proteins of Hendra virus was shown to induce protection against Nipah in African Green Monkeys.

In animals, especially in pigs, the virus causes a porcine respiratory and neurologic syndrome, locally known as "barking pig syndrome" or "one mile cough."

Ephrin B2 has been identified as the main receptor for the henipaviruses.

Figure 4 Structure of NIPAH VIRUS

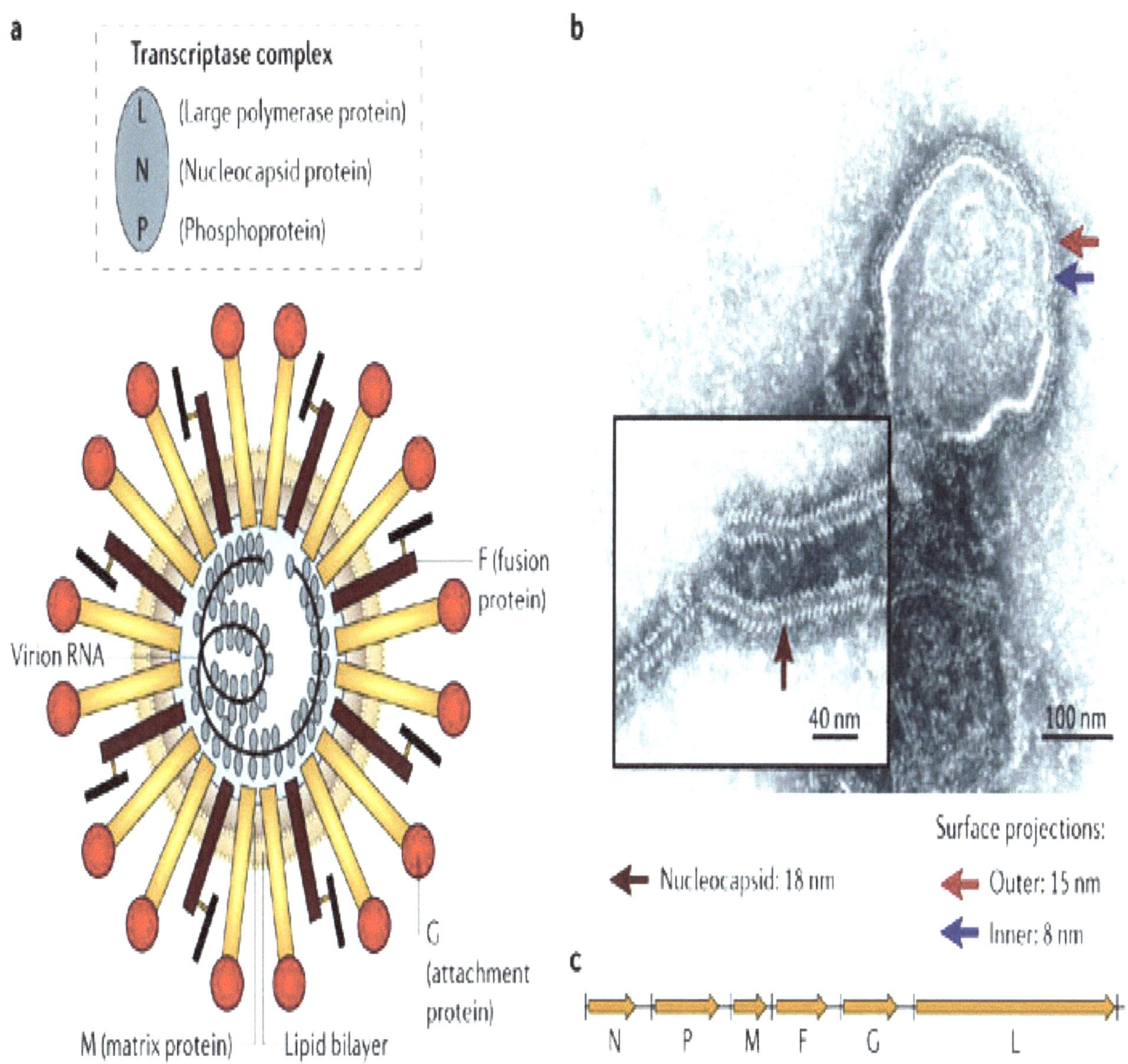

Figure 5 NIPAH virus in News

Nipah virus threat 'under control'

■ By Rina De Silva
rinade@nst.com.my

SUBANG JAYA: The Health Ministry has given an assurance that its Centre for Disease Control and Prevention is always on the alert for the Nipah virus.

"Our centre is advanced. In fact, we were the ones who identified the Nipah virus and we know how to control it," Health Minister Datuk Liow Tiong Lai said after launching the Health Day campaign organised by the Women Development Organisation of Malaysia and Puchong MCA division here yesterday.

"We are confident if there is an outbreak, we will know the source and be able to handle it. I don't think there is a threat here."

Liow said there has been no new cases of the Nipah virus in the country since the 1998-1999 outbreak when more than 100 people died.

None of the 649 encephalitis cases reported to the ministry from 2000 to October this year was caused by the Nipah virus.

The minister was commenting on the *New Sunday Times* front-page report "Nipah virus still a threat", which said that Universiti Malaya's department of medicine did not rule out the possibility of another Nipah outbreak in the country.

The NST quoted the department's consultant neurologist Dr Tan Chong Tin as saying the country was vulnerable as Bangladesh, Indonesia, Thailand, Cambodia and India have discovered bats infected with the Nipah virus.

Datuk Liow Tiong Lai says there is little evidence that foreigners pose a threat

Dr Tan had also said the movement of labour across borders could also pose a threat to Malaysia, home to migrant labour from these countries.

Liow said there was little evidence that the foreigners posed a threat. "The risk they carry is small or non-existent."

Discovery of Nipah virus a major breakthrough

IT WAS service to the country and he wasn't looking for fame or recognition. But Dr Tan Chong Tin who led the Nipah encephalitis investigation team from Universiti Malaya gained global acknowledgement for his work in discovering a new virus.

And last year, the team was recognised nationally when they were awarded joint recipients of the Merdeka Award in the health, science and technology category.

"We feel honoured to have contributed to the discovery. As human beings, it does feel good to have affirmation of our work – but we don't look for it," says Dr Tan humbly.

He also credits the actual discovery of the virus to Dr Chua Kaw Bing and says that "accepting the award was another way of recognising his discovery."

When asked how he feels about his team's accomplishment, he says it was a good piece of scientific work.

"It is widely recognised but more importantly, it made a difference to many people's lives. It also created real knowledge – not just locally, but internationally as well," he says, adding that their discovery had a great impact when there was an outbreak of the Nipah virus in Bangladesh and India.

However, he adds that there is another dimension apart from the recognition of the work done.

"This kind of award is quite timely as the country needs to recognise good scientific work. There is a need to put science and research in a higher hierarchy in our local values. This sort of emphasis will augur well for the future of our society and for professionalism in general."

Describing the early days of the outbreak, he says when the first cases surfaced in Ipoh, the health authorities thought it was Japanese encephalitis (JE) and health measures were taken accordingly.

However, it was not until cases appeared in Seremban three months later and patients were referred to the University Malaya Medical Centre (UMMC) that it came to their attention. But the realisation that the virus was completely unknown was not a *eureka* moment.

"Actually, the discovery of the disease was a gradual thing. Within the first three to four days, we knew that there was something about the virus that was different from JE. The evidence came cumulatively, and as more came in, we did tests to confirm it.

"We gradually began to feel more comfortable with it as evidence told us we were on the right track.

"Final confirmation came within 10 to 12 days and final identificatioin was done in the United States," he says.

Proud team: Dr Tan (far right) and Dr Chua (second from right) together with other members of the Nipah encephalitis investigation team from Universiti Malaya's Faculty of Medicine.

Figure 6 NIPAH VIRUS Outline

Description: Member of family Paramyxoviridae, single stranded RNA, negative sense, helical/enveloped, non-segmented. Fruit bats are believed to be the natural hosts, but human cases result from direct contact with infected pigs. Initially isolated following an outbreak of encephalitis in Singapore and Malaysia. It is closely related to Hendra virus and can infect several species.

Power:

Offenses

Attacks - Symptomatic cases begin with several days of fever, headache, myalgia, drowsiness, and confusion. Disease may progress to encephalitis coma, seizures, and suffocation.

Outcome - Many cases are asymptomatic and milder cases will progress tc full recovery. However, fifty percent of clinically apparent cases may die.

Speed - Illness begins with 3-14 days of fever and coma may occur within 24-48 hours. Total incubation period is generally between 4 and 18 days.

Defenses

Vaccines - none

Behavioral - Avoid contact with pigs in endemic areas.

Treatment - No drug therapies have yet been proven effective in treating Nipah infection. Intensive supportive care should be administered. Limited evidence suggests that Ribavirin may lessen the duration/severity of illness.

CHAPTER 6 RABIES

Introduction

Rabies (from Latin: *rabies*, "madness") is a viral disease that causes acute encephalitis in warm-blooded animals. The disease is zoonotic, meaning it can be transmitted to humans from another species (such as dogs), commonly by a bite from an infected animal. For a human, rabies is almost invariably fatal if postexposure prophylaxis is not administered prior to the onset of severe symptoms. The rabies virus infects the central nervous system, ultimately causing disease in the brain and death.

The rabies virus travels to the brain by following the peripheral nerves. The incubation period of the disease is usually a few months in humans, depending on the distance the virus must travel to reach the central nervous system. Once the rabies virus reaches the central nervous system and symptoms begin to show, the infection is virtually untreatable and usually fatal within days.

Early-stage symptoms of rabies are malaise, headache and fever, progressing to acute pain, violent movements, uncontrolled excitement, depression, and hydrophobia. Finally, the patient may experience periods of mania and lethargy, eventually leading to coma. The primary cause of death is usually respiratory insufficiency.

Rabies causes about 55,000 human deaths annually worldwide. 95% of human deaths due to rabies occur in Asia and Africa. Roughly 97% of human rabies cases result from dog bites. In the United States, animal control and vaccination programs have effectively eliminated domestic dogs as reservoirs of rabies. In several countries, including Australia and Japan, rabies carried by terrestrial animals has been eliminated entirely. While classical rabies has been eradicated in the United Kingdom, bats infected with a related virus have been found in the country on rare occasions

History

The term is derived from the Latin *rabies*, "madness". This, in turn, may be related to the Sanskrit *rabhas*, "to do violence". The Greeks derived the word *lyssa*, from *lud* or "violent"; this root is used in the name of the genus of rabies *Lyssavirus*

Because of its potentially violent nature, rabies has been known since *circa* 2000 B.C. The first written record of rabies is in the Mesopotamian Codex of Eshnunna (circa 1930 BC), which dictates that the owner of a dog showing symptoms of rabies should take preventive measure against bites. If another person were bitten by a rabid dog and later died, the owner was heavily fined

Rabies was considered a scourge for its prevalence in the 19th century. In France and Belgium, where Saint Hubert was venerated, the "St Hubert's Key" was heated and applied to cauterize the wound. By an application of magical thinking, dogs were branded with the key in hopes of protecting them from rabies. The fear of rabies was almost irrational, due to the significant number of vectors (mostly rabid dogs) and the absence of any efficacious treatment. It was not uncommon for a person, showing no signs of the disease, bitten by a dog merely suspected of being rabid, to commit suicide or to be killed by others.[66] This gave Louis Pasteur ample opportunity to test postexposure treatments from 1885.[69] In ancient medical times, the attachment of the tongue (the lingual frenulum, a mucous membrane) was cut and removed as this is where rabies was thought to originate. This practice ceased with the discovery of the actual cause of rabies

Virology

The rabies virus is the type species of the Lyssavirus genus, in the family *Rhabdoviridae*, order *Mononegavirales*. Lyssaviruses have helical symmetry, with a length of about 180 nm and a cross-section of about 75 nm These viruses are enveloped and have a single-stranded RNA genome with negative sense. The genetic information is packed as a ribonucleoprotein complex in which RNA is tightly bound by the viral nucleoprotein. The RNA genome of the virus encodes five genes whose order is highly conserved: nucleoprotein (N), phosphoprotein (P), matrix protein (M), glycoprotein (G), and the viral RNA polymerase (L).

Once within a muscle or nerve cell, the virus undergoes replication. The trimeric spikes on the exterior of the membrane of the virus interact with a specific cell receptor, the most likely one being the Acetylcholine receptor. The cellular membrane pinches in a procession known as pinocytosis and allows entry of the virus into the cell by way of an endosome. The virus then uses the acidic environment, which is necessary, of that endosome and binds to its membrane simultaneously, releasing its five proteins and single strand RNA into the cytoplasm. The L protein then transcribes five mRNA strands and a positive strand of RNA all from the original negative strand RNA using free nucleotides in the cytoplasm. These five mRNA strands are then translated into their corresponding proteins (P, L, N, G and M proteins) at free ribosomes in the cytoplasm. Some proteins require post-translative modifications. For example, the G protein travels through the rough endoplasmic reticulum, where it undergoes further folding, and is then transported to the Golgi apparatus, where a sugar group is added to it (glycosylation).

Where there are enough proteins, the viral polymerase will begin to synthesize new negative strands of RNA from the template of the positive strand RNA. These negative strands will then form complexes with the N, P, L and M proteins and then travel to the inner membrane of the cell, where a G protein has embedded itself in the membrane. The G protein then coils around the N-P-L-M complex of proteins taking some of the host cell membrane with it, which will form the new outer envelope of the virus particle. The virus then buds from the cell.

From the point of entry, the virus is neurotropic, traveling quickly along the neural pathways into the central nervous system, and then to other organs. The salivary glands receive high concentrations of the virus, thus allowing further transmission.

Figure 7 Rabies virus

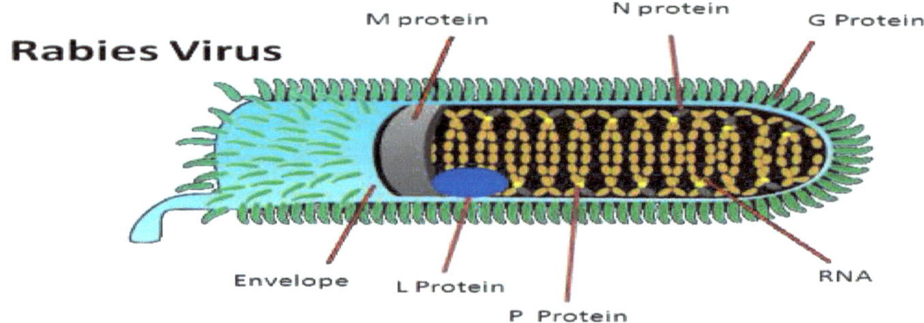

Figure 8 Rabies virus (TEM image)

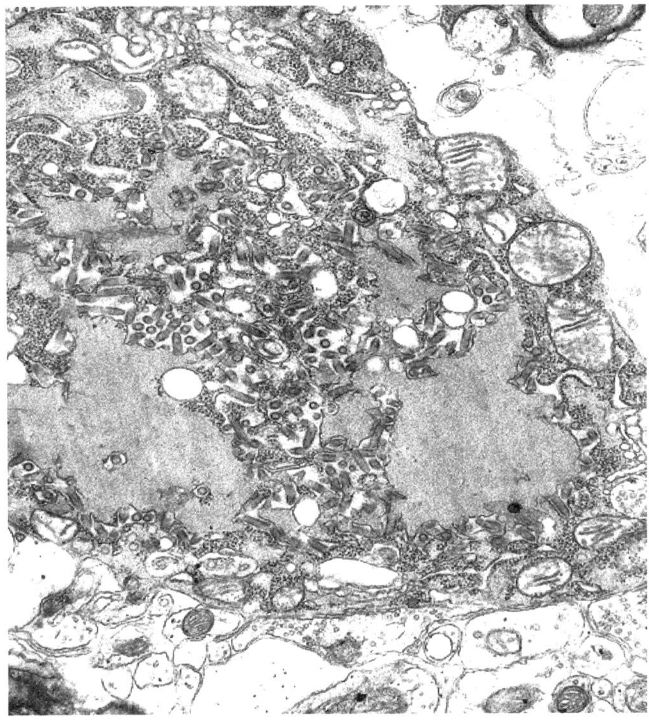

Transmission

Any warm-blooded animal, including humans, may become infected with the rabies virus and develop symptoms, although birds have only been known to be infected in experiments. The virus has even been adapted to grow in cells of poikilothermic ("cold-blooded") vertebrates. Most animals can be infected by the virus and can transmit the disease to humans. Infected bats, monkeys, raccoons, foxes, skunks, cattle, wolves, coyotes, dogs, mongooses (normally yellow mongoose) or cats present the greatest risk to humans.

Rabies may also spread through exposure to infected domestic farm animals, groundhogs, weasels, bears, raccoons, skunks and other wild carnivorans.

Small rodents, such as squirrels, hamsters, guinea pigs, gerbils, chipmunks, rats, and mice, and lagomorphs such as rabbits and hares, are almost never found to be infected with rabies and are not known to transmit rabies to humans. The Virginia opossum is resistant but not immune to rabies.

The virus is usually present in the nerves and saliva of a symptomatic rabid animal. The route of infection is usually, but not always, by a bite. In many cases, the infected animal is exceptionally aggressive, may attack without provocation, and exhibits otherwise uncharacteristic behavior. This is an example of a viral pathogen modifying the behavior of its host to facilitate its transmission to other hosts.

Transmission between humans is extremely rare. A few cases have been recorded through transplant surgery. After a typical human infection by bite, the virus enters the peripheral nervous system. It then travels along the afferent nerves toward the central nervous system .During this phase, the virus cannot be easily detected within the host, and vaccination may still confer cell-mediated immunity to prevent symptomatic rabies. When the virus reaches the brain, it rapidly causes encephalitis, the prodromal phase, and is the beginning of the symptoms. Once the patient becomes symptomatic, treatment is almost never effective and mortality is over 99%. Rabies may also inflame the spinal cord, producing transverse myelitis

Epidemiology

In 2010, an estimated 26,000 people died from rabies, down from 54,000 in 1990.[1] The majority of the deaths occurred in Asia and Africa India has the highest rate of human rabies in the world, primarily because of stray dogs whose number has greatly increased since a 2001 law forbade the killing of dogs. Effective control and treatment of rabies in India is also hindered by a form of mass hysteria or group delusion known as puppy pregnancy syndrome (PPS). Dog bite victims with PPS (both male and female) become convinced that puppies are growing inside them, and often seek help from faith healers rather than from conventional medical services. In cases where the bite was from a rabid dog, this decision can prove fatal. Dr. Nitai Kishore Marik, former district medical officer of West Midnapur, states "I have seen scores of cases of rabies that reached our hospitals very late because of the intervention of faith healers. We could not save those lives." An estimated 20,000 people die every year from rabies in India — more than a third of the global toll. As of 2007, Vietnam had the second-highest rate, followed by Thailand; in these countries, the virus is primarily transmitted through canines (feral dogs and other wild canine species). Another source of rabies in Asia is the pet boom. In 2006 China introduced the "one-dog policy" in Beijing to control the problem.

The rabies virus survives in widespread, varied, rural fauna reservoirs. It is present in the animal populations of almost every country in the world except Australia and New Zealand Australian bat lyssavirus (ABLV), discovered in 1996, is similar to rabies and is believed to be prevalent in native bat populations. In some countries, such as those in western Europe and Oceania, rabies is considered to be prevalent among bat populations only.

In Asia and in parts of the Americas and Africa, dogs remain the principal host. Mandatory vaccination of animals is less effective in rural areas. Especially in developing countries, pets may not be privately kept and their destruction may be unacceptable. Oral vaccines can be safely distributed in baits, a practice that has successfully reduced rabies in rural areas of Canada, France, and the United States. In Montréal, Quebec, Canada, baits are successfully used on raccoons in the Mont-Royal Park area. Vaccination campaigns may be expensive, and cost-benefit analysis suggests baits may be a cost-effective method of control. In Ontario, a dramatic drop in rabies was recorded when an aerial bait-vaccination campaign was launched.

Rabies is common among wild animals in the US. Bats, raccoons, skunks and foxes account for almost all reported cases (98% in 2009). Rabid bats are found in all 48 contiguous states. Other reservoirs are more limited geographically; for example, the raccoon rabies virus variant is only found in a relatively narrow band along the East Coast. Due to a high public awareness of the virus, efforts at vaccination of domestic animals and curtailment of feral populations, and availability of post exposure prophylaxis, incidents of rabies in humans are very rare. A total of 49 cases of the disease was reported in the country between 1995 and 2011; of these, 11 are thought to have been acquired abroad. Almost all domestically acquired cases are attributed to bat bites

In Switzerland, the disease has been virtually eradicated after scientists placed chicken heads laced with live attenuated vaccine in the Swiss Alps. The foxes of Switzerland, proven to be the main source of rabies in the country, ate the chicken heads and immunized themselves.

Diagnosis

Rabies can be difficult to diagnose because, in the early stages, it is easily confused with other diseases or aggressiveness.

The reference method for diagnosing rabies is the Fluorescent Antibody Test (FAT) which is recommended by World Health organization (WHO). The FAT relies on the ability of a detector molecule (usually fluorescein isothiocyanate) coupled with a rabies specific antibody forming a conjugate to bind to and allow the visualisation of rabies antigen using fluorescent microscopy techniques. Microscopic analysis of samples is the only direct method that allows for the identification of rabies virus-specific antigen in a short time and at a reduced cost, irrespective of geographical origin and status of the host. It has to be regarded as the first step in diagnostic procedures for all laboratories. Autolysed samples can, however, reduce the sensitivity and specificity of the FAT The RT PCR assays proved to be a sensitive and specific tool for routine diagnostic purposes, particularly in decomposed samples or archival specimens. The diagnosis can be reliably made from brain samples taken after death. The diagnosis can also be made from saliva, urine, and cerebrospinal fluid samples, but this is not as sensitive and reliable as brain samples. Cerebral inclusion bodies called Negri bodies are 100% diagnostic for rabies infection but are found in only about 80% of cases. If possible, the animal from which the bite was received should also be examined for rabies

The differential diagnosis in a case of suspected human rabies may initially include any cause of encephalitis, in particular infection with viruses such as herpesviruses, enteroviruses, and arboviruses such as West Nile virus. The most important viruses to rule out are herpes simplex virus type one, varicella zoster virus, and (less commonly) enteroviruses,

New causes of viral encephalitis are also possible, as was evidenced by the 1999 outbreak in Malaysia of 300 cases of encephalitis with a mortality rate of 40% caused by Nipah virus, a newly recognized paramyxovirus Likewise, well-known viruses may be introduced into new locales, as is illustrated by the recent outbreak of encephalitis due to West Nile virus in the eastern United States. Epidemiologic factors, such as season, geographic location, and the patient's age, travel history, and possible exposure to bites, rodents, and ticks, may help direct the diagnosis.

Cheaper rabies diagnosis will become possible for low-income settings: accurate rabies diagnosis can be done at a tenth of the cost of traditional testing using basic light microscopy techniques.

All human cases of rabies were fatal until a vaccine was developed in 1885 by Louis Pasteur and Émile Roux. Their original vaccine was harvested from infected rabbits, from which the virus in the nerve tissue was weakened by allowing it to dry for five to 10 days.[1] Similar nerve tissue-derived vaccines are still used in some countries, as they are much cheaper than modern cell culture vaccines.

The human diploid cell rabies vaccine was started in 1967; a new and less expensive purified chicken embryo cell vaccine and purified vero cell rabies vaccine are now available. A recombinant vaccine called V-RG has been successfully used in Belgium, France, Germany, and the United States to prevent outbreaks of rabies in undomesticated animals. Currently, immunization prior to exposure has been used in both human and nonhuman populations, where, as in many jurisdictions, domesticated animals are required to be vaccinated.

In the United States, since the widespread vaccination of domestic dogs and cats and the development of effective human vaccines and immunoglobulin treatments, the number of recorded human deaths from rabies has dropped from 100 or more annually in the early 20th century to one to two per year, mostly caused by bat bites, which may go unnoted by the victim and hence untreated.

The Missouri Department of Health and Senior Services Communicable Disease Surveillance 2007 Annual Report states the following can help reduce the risk of contracting rabies

- Vaccinating dogs, cats, rabbits, and ferrets against rabies
- Keeping pets under supervision
- Not handling wild animals or strays
- Contacting an animal control officer upon observing a wild animal or a stray, especially if the animal is acting strangely
- If bitten by an animal, washing the wound with soap and water for 10 to 15 minutes and contacting a healthcare provider to determine if post-exposure prophylaxis is required

September 28 is World Rabies Day, which promotes the information, prevention, and elimination of the disease

Treatment

Treatment after exposure is highly successful in preventing the disease if administered promptly, in general within 10 days of infection. Thoroughly washing the wound as soon as possible with soap and water for approximately five minutes is very effective in reducing the number of viral particles. "If available, a virucidal antiseptic such as povidone-iodine, iodine tincture, aqueous iodine solution, or alcohol (ethanol) should be applied after washing. Exposed mucous membranes such as eyes, nose or mouth should be flushed well with water."

In the US, the Centers for Disease Control and Prevention recommends patients receive one dose of human rabies immunoglobulin (HRIG) and four doses of rabies vaccine over a 14-day period. The immunoglobulin dose should not exceed 20 units per kilogram body weight. HRIG is expensive and constitutes the vast majority of the cost of postexposure treatment, ranging as high as several thousand dollars. As much as possible of this dose should be infiltrated around the bites, with the remainder being given by deep intramuscular injection at a site distant from the vaccination site. The first dose of rabies vaccine is given as soon as possible after exposure, with additional doses on days three, seven and 14 after the first. Patients who have previously received pre-exposure vaccination do not receive the immunoglobulin, only the postexposure vaccinations on days 0 and 2.

Modern cell-based vaccines are similar to flu shots in terms of pain and side effects. The old nerve-tissue-based vaccinations that require multiple painful injections into the abdomen with a large needle are inexpensive, but are being phased out and replaced by affordable World Health Organization intradermal vaccination regimens.

Intramuscular vaccination should be given into the deltoid, not gluteal area, which has been associated with vaccination failure due to injection into fat rather than muscle. In infants, the lateral thigh is used as for routine childhood vaccinations.

Awakening to find a bat in the room, or finding a bat in the room of a previously unattended child or mentally disabled or intoxicated person, is regarded as an indication for postexposure prophylaxis (PEP). The recommendation for the precautionary use of PEP in occult bat encounters where no contact is recognized has been questioned in the medical literature, based on a cost-benefit analysis. However, a 2002 study has supported the protocol of precautionary administering of PEP where a child or mentally compromised individual has been alone with a bat, especially in sleep areas, where a bite or exposure may occur without the victim being aware. Begun with little or no delay, PEP is 100% effective against rabies. In the case in which there has been a significant delay in administering PEP, the treatment should be administered regardless, as it may still be effective

In 2004, American teenager Jeanna Giese survived an infection of rabies unvaccinated. She was placed into an induced coma upon onset of symptoms and given ketamine, midazolam, ribavirin, and amantadine. Her doctors administered treatment based on the hypothesis that detrimental effects of rabies were caused by temporary dysfunctions in the brain and could be avoided by inducing a temporary partial halt in brain function that would protect the brain from damage while giving the immune system time to defeat the virus. After 31 days of isolation and 76 days of hospitalization, Giese was released from the hospital. She survived with all higher level brain functions, but an inability to walk and balance. On a podcast of NPR's Radiolab, Giese recounted, "I had to learn how to stand and then to walk, turn around, move my toes. I was really, after rabies, a new born baby who couldn't do anything. I had to relearn that all...mentally I knew how to do stuff but my body wouldn't cooperate with what I wanted it to do. It definitely took a toll on me psychologically. You know I'm still recovering. I'm not completely back. Stuff like balance and, um, I can't run normally." Giese's treatment regimen became known as the "Milwaukee protocol", which has since undergone revision with the second version omitting the use of ribavirin. Two of 25 patients survived when treated under the first protocol. A further 10 patients have been treated under the revised protocol, with a further two survivors. The anesthetic drug ketamine has shown the potential for rabies virus inhibition in rats, and is used as part of the Milwaukee protocol.

On April 10, 2008, in Cali, Colombia, a boy of 11 was reported to have survived rabies and the induced coma without noticeable brain damage. On June 12, 2011, Precious Reynolds, an eight-year-old girl from Humboldt County, California, became the third reported person in the world and the second in the United States to have recovered from rabies without receiving PEP.

Images related to Rabies

Figure 9 Rabies virus Transmission

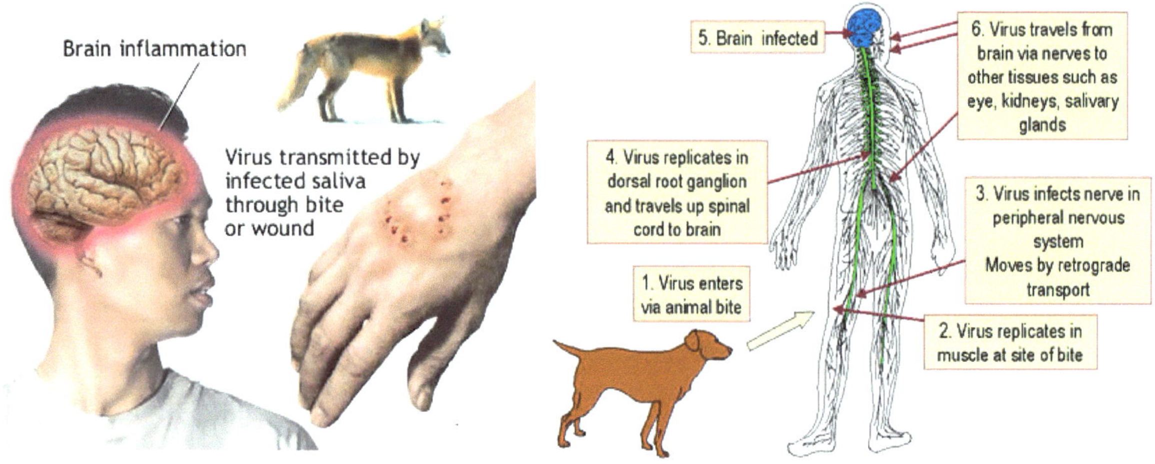

Figure 10 Rabies virus Outline

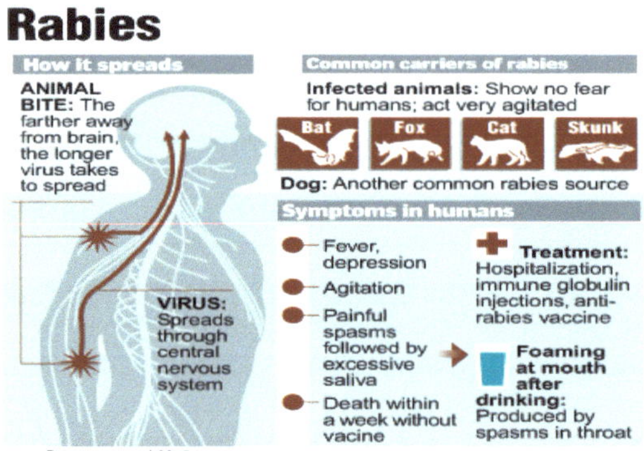

Figure 11 Rabies virus Treatment outline

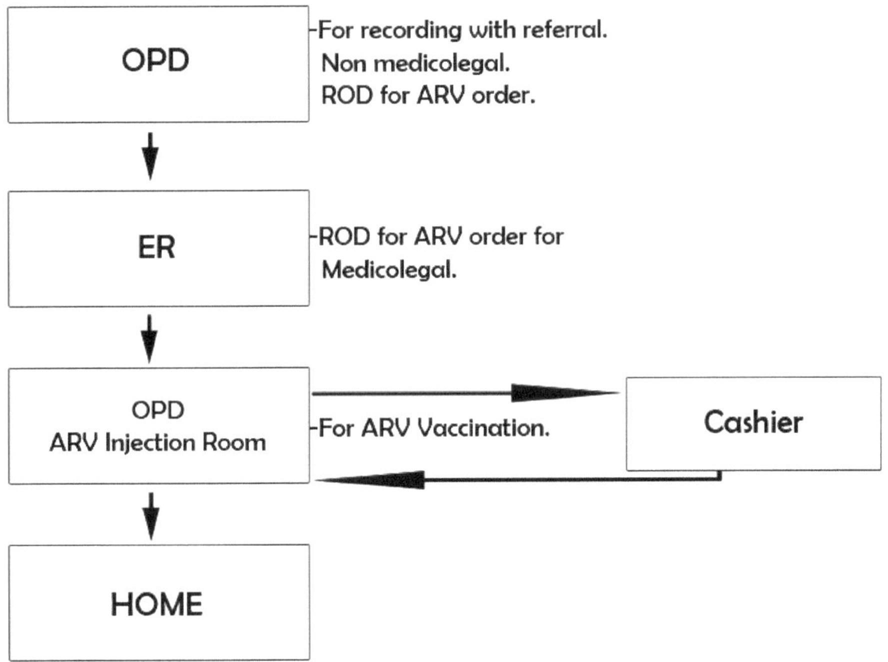

Figure 12 Patients suffering from RABIES

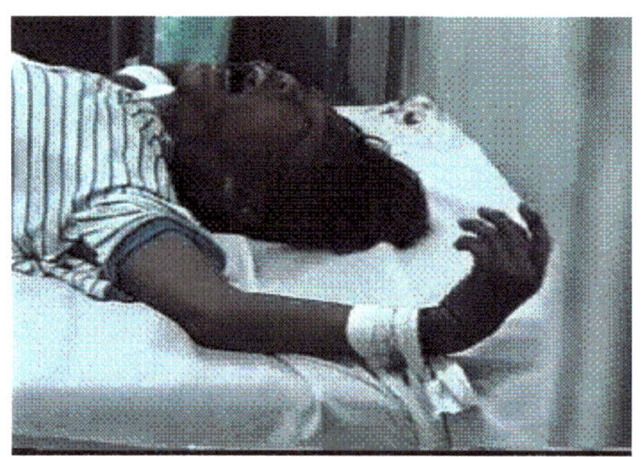

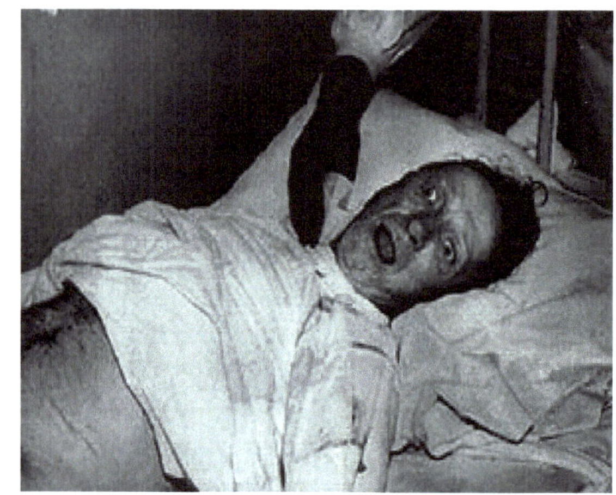

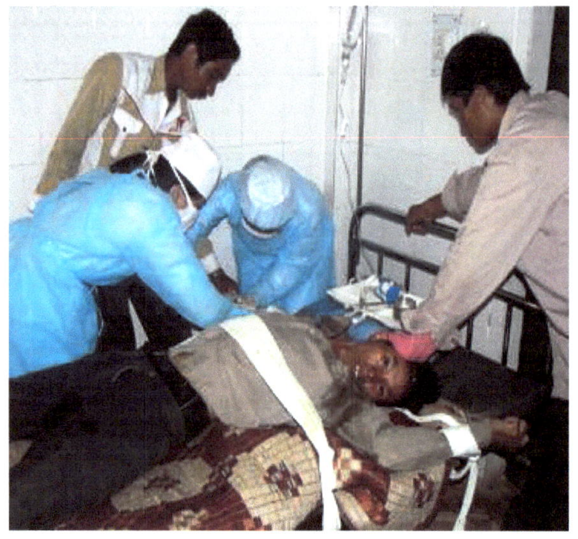

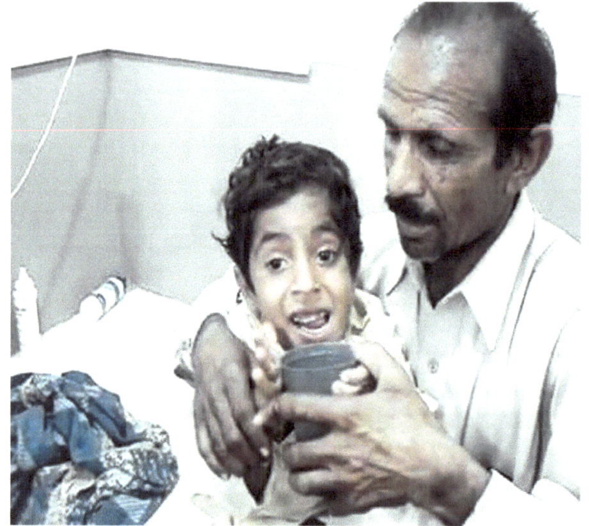

CHAPTER 7 H5N1 VIRUS

Influenza A virus subtype H5N1, also known as "bird flu", A(H5N1) or simply H5N1, is a subtype of the influenza A virus which can cause illness in humans and many other animal species. A bird-adapted strain of H5N1, called HPAI A(H5N1) for "highly pathogenic avian influenza virus of type A of subtype H5N1", is the causative agent of H5N1 flu, commonly known as "avian influenza" or "bird flu". It is enzootic in many bird populations, especially in Southeast Asia. One strain of HPAI A(H5N1) is spreading globally after first appearing in Asia. It is epizootic (an epidemic in nonhumans) and panzootic (affecting animals of many species, especially over a wide area), killing tens of millions of birds and spurring the culling of hundreds of millions of others to stem its spread. Most references to "bird flu" and H5N1 in the popular media refer to this strain.

According to the FAO Avian Influenza Disease Emergency Situation Update, H5N1 pathogenicity is gradually continuing to rise in endemic areas, but the avian influenza disease situation in farmed birds is being held in check by vaccination. Eleven outbreaks of H5N1 were reported worldwide in June 2008 in five countries (China, Egypt, Indonesia, Pakistan and Vietnam) compared to 65 outbreaks in June 2006 and 55 in June 2007. The "global HPAI situation can be said to have improved markedly in the first half of 2008 [but] cases of HPAI are still underestimated and underreported in many countries because of limitations in country disease surveillance systems".[3] In July 2013 the WHO announced a total of 630 *confirmed* human cases which resulted in the deaths of 375 people since 2003.[4]

A filtered and purified influenza A vaccine for humans is being developed, and many countries have recommended it be stockpiled so, if an avian influenza pandemic starts jumping to humans, the vaccine can quickly be administered to avoid loss of life.

Research has shown that a highly contagious strain of H5N1, one that might allow airborne transmission between mammals, can be reached in only a few mutations, raising the spectre of a pandemic epidemy or even bioterrorism[5] in the human population

HPAI A(H5N1) is considered an avian disease, although there is some evidence of limited human-to-human transmission of the virus. A risk factor for contracting the virus is handling of infected poultry, but transmission of the virus from infected birds to humans is inefficient. Still, around 60% of humans known to have been infected with the current Asian strain of HPAI A(H5N1) have died from it, and H5N1 may mutate or reassort into a strain capable of efficient human-to-human transmission. In 2003, world-renowned virologist Robert G. Webster published an article titled "The world is teetering on the edge of a pandemic that could kill a large fraction of the human population" in *American Scientist*. He called for adequate resources to fight what he sees as a major world threat to possibly billions of lives.[8] On September 29, 2005, David Nabarro, the newly appointed Senior United Nations System Coordinator for Avian and Human Influenza, warned the world that an outbreak of avian influenza could kill anywhere between 5 million and 150 million people.[9] Experts have identified key events (creating new clades, infecting new species, spreading to new areas) marking the progression of an avian flu virus towards becoming pandemic, and many of those key events have occurred more rapidly than expected.

Due to the high lethality and virulence of HPAI A(H5N1), its endemic presence, its increasingly large host reservoir, and its significant ongoing mutations, the H5N1 virus is the world's largest current pandemic threat, and billions of dollars are being spent researching H5N1 and preparing for a potential influenza pandemic.[10] At least 12 companies and 17 governments are developing prepandemic influenza vaccines in 28 different clinical trials that, if successful, could turn a deadly pandemic infection into a nondeadly one. Full-scale production of a vaccine that could prevent any illness at all from the strain would require at least three months after the virus's emergence to begin, but it is hoped that vaccine production could increase until one billion doses were produced by one year after the initial identification of the virus.[11]

H5N1 may cause more than one influenza pandemic, as it is expected to continue mutating in birds regardless of whether humans develop herd immunity to a future pandemic strain.[12] Influenza pandemics from its genetic offspring may include influenza A virus subtypes other than H5N1.[13] While genetic analysis of the H5N1 virus shows that influenza pandemics from its genetic offspring can easily be far more lethal than the Spanish flu pandemic,[14] planning for a future influenza pandemic is based on what can be done and there is no higher Pandemic Severity Index level than a Category 5 pandemic which, roughly speaking, is any pandemic as bad as the Spanish flu or worse; and for which *all*intervention measures are to be used

The avian influenza hemagglutinin binds alpha 2-3 sialic acid receptors, while human influenza hemagglutinins bind alpha 2-6 sialic acid receptors.[17]This means when the H5N1 strain infects humans, it will replicate in the lower respiratory tract, and consequently will cause viral pneumonia. There is as yet no human form of H5N1, so all humans who have caught it so far have caught **avian** H5N1. In general, humans who catch a humanized influenza A virus (a human flu virus of type A) usually have symptoms that include fever, cough, sore throat, muscle aches, conjunctivitis, and, in severe cases, breathing problems and pneumonia that may be fatal. The severity of the infection depends in large part on the state of the infected persons'immune systems and whether they had been exposed to the strain before (in which case they would be partially immune). No one knows if these or other symptoms will be the symptoms of a humanized H5N1 flu.

The reported mortality rate of highly pathogenic H5N1 avian influenza in a human is high; WHO data indicate 60% of cases classified as H5N1 resulted in death. However, there is some evidence the actual mortality rate of avian flu could be much lower, as there may be many people with milder symptoms who do not seek treatment and are not counted.[19][20]

In one case, a boy with H5N1 experienced diarrhea followed rapidly by a coma without developing respiratory or flu-like symptoms. There have been studies of the levels of cytokines in humans infected by the H5N1 flu virus. Of particular concern is elevated levels of tumor necrosis factor-alpha, a protein associated with tissue destruction at sites of infection and increased production of other cytokines. Flu virus-induced increases in the level of cytokines is also associated with flu symptoms, including fever, chills, vomiting and headache. Tissue damage associated with pathogenic flu virus infection can ultimately result in death.[8] The inflammatory cascade triggered by H5N1 has been called a 'cytokine storm' by some, because of what seems to be a positive feedback process of damage to the body resulting from immune system stimulation. H5N1 induces higher levels of cytokines than the more common flu virus types

The first known strain of HPAI A(H5N1) (called A/chicken/Scotland/59) killed two flocks of chickens in Scotland in 1959, but that strain was very different from the current highly pathogenic strain of H5N1. The dominant strain of HPAI A(H5N1) in 2004 evolved from 1999 to 2002 creating the Z genotype.[23] It has also been called "Asian lineage HPAI A(H5N1)".

Asian lineage HPAI A(H5N1) is divided into two antigenic clades. "Clade 1 includes human and bird isolates from Vietnam, Thailand, and Cambodia and bird isolates from Laos and Malaysia. Clade 2 viruses were first identified in bird isolates from China, Indonesia, Japan, and South Korea before spreading westward to the Middle East, Europe, and Africa. The clade 2 viruses have been primarily responsible for human H5N1 infections that have occurred during late 2005 and 2006, according to WHO. Genetic analysis has identified six subclades of clade 2, three of which have a distinct geographic distribution and have been implicated in human infections: Map

- Subclade 1, Indonesia
- Subclade 2, Europe, Middle East, and Africa (called EMA)
- Subclade 3, China"

A 2007 study focused on the EMA subclade has shed further light on the EMA mutations. "The 36 new isolates reported here greatly expand the amount of whole-genome sequence data available from recent avian influenza (H5N1) isolates. Before our project, GenBank contained only 5 other complete genomes from Europe for the 2004–2006 period, and it contained no whole genomes from the Middle East or northern Africa. Our analysis showed several new findings. First, all European, Middle Eastern, and African samples fall into a clade that is distinct from other contemporary Asian clades, all of which share common ancestry with the original 1997 Hong Kong strain. Phylogenetic trees built on each of the 8 segments show a consistent picture of 3 lineages, as illustrated by the HA tree shown in Figure 1. Two of the clades contain exclusively Vietnamese isolates; the smaller of these, with 5 isolates, we label V1; the larger clade, with 9 isolates, is V2. The remaining 22 isolates all fall into a third, clearly distinct clade, labeled EMA, which comprises samples from Europe, the Middle East, and Africa. Trees for the other 7 segments display a similar topology, with clades V1, V2, and EMA clearly separated in each case. Analyses of all available complete influenza (H5N1) genomes and of 589 HA sequences placed the EMA clade as distinct from the major clades circulating in People's Republic of China, Indonesia, and Southeast Asia

Terminology

H5N1 isolates are identified like this actual HPAI A(H5N1) example, **A/chicken/Nakorn-Patom/Thailand/CU-K2/04(H5N1)**:

- **A** stands for the genus of influenza (A, B or C). **chicken** is the animal species the isolate was found in (note: human isolates lack this component term and are thus identified as human isolates by default)
- **Nakorn-Patom/Thailand** is the place this specific virus was isolated
- **CU-K2** is the laboratory reference number that identifies it from other influenza viruses isolated at the same place and year
- **04** represents the year of isolation 2004
- **H5** stands for the fifth of several known types of the protein hemagglutinin.
- **N1** stands for the first of several known types of the protein neuraminidase.

Other examples include: A/duck/Hong Kong/308/78(H5N3), A/avian/NY/01(H5N2), A/chicken/Mexico/31381-3/94(H5N2), and A/shoveler/Egypt/03(H5N2).[27]

As with other avian flu viruses, H5N1 has strains called "highly pathogenic" (HP) and "low-pathogenic" (LP). Avian influenza viruses that cause HPAI are highly virulent, and mortality rates in infected flocks often approach 100%. LPAI viruses have negligible virulence, but these viruses can serve as progenitors to HPAI viruses. The current strain of H5N1 responsible for the deaths of birds across the world is an HPAI strain; all other current strains of H5N1, including a North American strain that causes no disease at all in any species, are LPAI strains. All HPAI strains identified to date have involved H5 and H7 subtypes. The distinction concerns pathogenicity in poultry, not humans. Normally, a highly pathogenic avian virus is not highly pathogenic to either humans or nonpoultry birds. This current deadly strain of H5N1 is unusual in being deadly to so many species, including some, like domestic cats, never previously susceptible to any influenza virus.[28]

Genetic structure and related subtypes

H5N1 is a subtype of the species *Influenza A virus* of the *Influenza virus A* genus of the *Orthomyxoviridae* family. Like all other influenza A subtypes, the H5N1 subtype is an RNA virus. It has a segmented genome of eight negative sense, single-strands of RNA, abbreviated as PB2, PB1, PA, HA, NP, NA, MP and NS.

HA codes for hemagglutinin, an antigenic glycoprotein found on the surface of the influenza viruses and is responsible for binding the virus to the cell that is being infected. NA codes for neuraminidase, an antigenic glycosylated enzyme found on the surface of the influenza viruses. It facilitates the release of progeny viruses from infected cells. The hemagglutinin (HA) and neuraminidase (NA) RNA strands specify the structure of proteins that are most medically relevant as targets for antiviral drugs and antibodies. HA and NA are also used as the basis for the naming of the different subtypes of influenza A viruses. This is where the *H* and *N*come from in *H5N1*.

Influenza A viruses are significant for their potential for disease and death in humans and other animals. Influenza A virus subtypes that have been confirmed in humans, in order of the number of known human pandemic deaths that they have caused, include:

- H1N1, which caused the 1918 flu pandemic ("Spanish flu") and currently is causing seasonal human flu and the 2009 flu pandemic ("swine flu")
- H2N2, which caused "Asian flu"
- H3N2, which caused "Hong Kong flu" and currently causes seasonal human flu
- H5N1, ("bird flu"), which is noted for having a strain (Asian-lineage HPAI H5N1) that kills over half the humans it infects, infecting and killing species that were never known to suffer from influenza viruses before (e.g. cats), being unable to be stopped by culling all involved poultry - some think due to being endemic in wild birds, and causing billions of dollars to be spent in flu pandemic preparation and preventiveness
- H7N7, which has unusual zoonotic potential and killed one person
- H1N2, which is currently endemic in humans and pigs and causes seasonal human flu
- H9N2, which has infected three people
- H7N2, which has infected two people
- H7N3, which has infected two people

- H10N7, which has infected two people
- H7N9, which in 2013 (as of May) has caused 131 cases including 31 deaths

Low pathogenic H5N1

Low pathogenic avian influenza H5N1 (LPAI H5N1) also called "North American" H5N1 commonly occurs in wild birds. In most cases, it causes minor sickness or no noticeable signs of disease in birds. It is not known to affect humans at all. The only concern about it is that it is possible for it to be transmitted to poultry and in poultry mutate into a highly pathogenic strain.

- 1966 - LPAI H5N1 A/Turkey/Ontario/6613/1966(H5N1) was detected in a flock of infected turkeys in Ontario, Canada
- 1975 – LPAI H5N1 was detected in a wild mallard duck and a wild blue goose in Wisconsin.
- 1981 and 1985 – LPAI H5N1 was detected in ducks by the University of Minnesota conducting a sampling procedure in which sentinel ducks were monitored in cages placed in the wild for a short period of time.
- 1983 – LPAI H5N1 was detected in ring-billed gulls in Pennsylvania.
- 1986 - LPAI H5N1 was detected in a wild mallard duck in Ohio.
- 2005 - LPAI H5N1 was detected in ducks in Manitoba, Canada.
- 2008 - LPAI H5N1 was detected in ducks in New Zealand.
- 2009 - LPAI H5N1 was detected in commercial poultry in British Columbia.[32]

"In the past, there was no requirement for reporting or tracking LPAI H5 or H7 detections in wild birds so states and universities tested wild bird samples independently of USDA. Because of this, the above list of previous detections might not be all inclusive of past LPAI H5N1 detections. However, the World Organization for Animal Health (OIE) recently changed its requirement of reporting detections of avian influenza. Effective in 2006, all confirmed LPAI H5 and H7 AI subtypes must be reported to the OIE because of their potential to mutate into highly pathogenic strains. Therefore, USDA now tracks these detections in wild birds, backyard flocks, commercial flocks and live bird markets

High mutation rate

Influenza viruses have a relatively high mutation rate that is characteristic of RNA viruses. The segmentation of its genome facilitates genetic recombination by segment reassortment in hosts infected with two different influenza viruses at the same time. A previously uncontagious strain may then be able to pass between humans, one of several possible paths to a pandemic.

The ability of various influenza strains to show species-selectivity is largely due to variation in the hemagglutinin genes. Genetic mutations in the hemagglutinin gene that cause single amino acidsubstitutions can significantly alter the ability of viral hemagglutinin proteins to bind to receptors on the surface of host cells. Such mutations in avian H5N1 viruses can change virus strains from being inefficient at infecting human cells to being as efficient in causing human infections as more common human influenza virus types. This doesn't mean that one amino acid substitution can cause a pandemic, but it does mean that one amino acid substitution can cause an avian flu virus that is not pathogenic in humans to become pathogenic in humans.

Influenza A virus subtype H3N2 is endemic in pigs in China, and has been detected in pigs in Vietnam, increasing fears of the emergence of new variant strains. The dominant strain of annual flu virus in January 2006 was H3N2, which is now resistant to the standard antiviral drugs amantadine and rimantadine. The possibility of H5N1 and H3N2 exchanging genes through reassortment is a major concern. If a reassortment in H5N1 occurs, it might remain an H5N1 subtype, or it could shift subtypes, as H2N2 did when it evolved into the Hong Kong Flu strain of H3N2.

Both the H2N2 and H3N2 pandemic strains contained avian influenza virus RNA segments. "While the pandemic human influenza viruses of 1957 (H2N2) and 1968 (H3N2) clearly arose through reassortment between human and avian viruses, the influenza virus causing the 'Spanish flu' in 1918 appears to be entirely derived from an avian source".

Prevention

Vaccine

There are several H5N1 vaccines for several of the avian H5N1 varieties, but the continual mutation of H5N1 renders them of limited use to date: while vaccines can sometimes provide cross-protection against related flu strains, the best protection would be from a vaccine specifically produced for any future pandemic flu virus strain. Dr. Daniel Lucey, co-director of the Biohazardous Threats and Emerging Diseases graduate program at Georgetown University has made this point, "There is no H5N1 pandemic so there can be no pandemic vaccine". However, "pre-pandemic vaccines" have been created; are being refined and tested; and do have some promise both in furthering research and preparedness for the next pandemic. Vaccine manufacturing companies are being encouraged to increase capacity so that if a pandemic vaccine is needed, facilities will be available for rapid production of large amounts of a vaccine specific to a new pandemic strain.

Treatment

There is no highly effective treatment for H5N1 flu, but oseltamivir (commercially marketed by Roche as Tamiflu), can sometimes inhibit the influenza virus from spreading inside the user's body. This drug has become a focus for some governments and organizations trying to prepare for a possible H5N1 pandemic. On April 20, 2006, Roche AG announced that a stockpile of three million treatment courses of Tamiflu are waiting at the disposal of the World Health Organization to be used in case of a flu pandemic; separately Roche donated two million courses to the WHO for use in developing nations that may be affected by such a pandemic but lack the ability to purchase large quantities of the drug.

However, WHO expert Hassan al-Bushra has said:

"Even now, we remain unsure about Tamiflu's real effectiveness. As for a vaccine, work cannot start on it until the emergence of a new virus, and we predict it would take six to nine months to develop it. For the moment, we cannot by any means count on a potential vaccine to prevent the spread of a contagious influenza virus, whose various precedents in the past 90 years have been highly pathogenic".

Animal and lab studies suggest that Relenza (zanamivir), which is in the same class of drugs as Tamiflu, may also be effective against H5N1. In a study performed on mice in 2000, "zanamivir was shown to be efficacious in treating avian influenza viruses H9N2, H6N1, and H5N1 transmissible to mammals".In addition, mice studies suggest the combination of zanamivir, celecoxib and mesalazine looks promising producing a 50% survival rate compared to no survival in the placebo arm. While no one knows if zanamivir will be useful or not on a yet to exist pandemic strain of H5N1, it might be useful to stockpile zanamivir as well as oseltamivir in the event of an H5N1 influenza pandemic. Neither oseltamivir nor zanamivir can currently be manufactured in quantities that would be meaningful once efficient human transmission starts. In September, 2006, a WHO scientist announced that studies had confirmed cases of H5N1 strains resistant to Tamiflu and Amantadine. Tamiflu-resistant strains have also appeared in the EU, which remain sensitive to Relenza.[contact with surfaces contaminated with them. H5N1 remains infectious after over 30 days at 0°C (32.0°F) (over one month at freezing temperature) or 6 days at 37°C (98.6°F) (one week at human body temperature); at ordinary temperatures it lasts in the environment for weeks. In Arctic temperatures, it does not degrade at all.

Because migratory birds are among the carriers of the highly pathogenic H5N1 virus, it is spreading to all parts of the world. H5N1 is different from all previously known highly pathogenic avian flu viruses in its ability to be spread by animals other than poultry.

In October 2004, researchers discovered H5N1 is far more dangerous than was previously believed. Waterfowl were revealed to be directly spreading this highly pathogenic strain to chickens, crows, pigeons, and other birds, and the virus was increasing its ability to infect mammals, as well. From this point on, avian flu experts increasingly referred to containment as a strategy that can delay, but not ultimately prevent, a future avian flu pandemic.

"Since 1997, studies of influenza A (H5N1) indicate that these viruses continue to evolve, with changes in antigenicity and internal gene constellations; an expanded host range in avian species and the ability to infect felids; enhanced pathogenicity in experimentally infected mice and ferrets, in which they cause systemic infections; and increased environmental stability."

The New York Times, in an article on transmission of H5N1 through smuggled birds, reports Wade Hagemeijer of Wetlands International stating, "We believe it is spread by both bird migration and trade, but that trade, particularly illegal trade, is more important"

On September 27, 2007 researchers reported the H5N1 bird flu virus can also pass through a pregnant woman's placenta to infect the fetus. They also found evidence of what doctors had long suspected — the virus not only affects the lungs, but also passes throughout the body into the gastrointestinal tract, the brain, liver, and blood cells

Images related to H5N1

Figure 13 H5N1 Virus

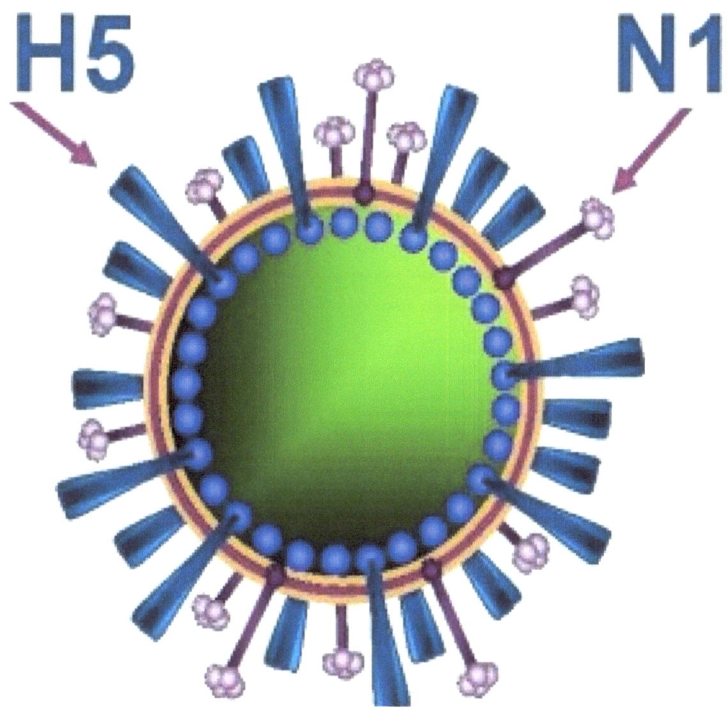

Figure 14 H5N1 Vaccine production

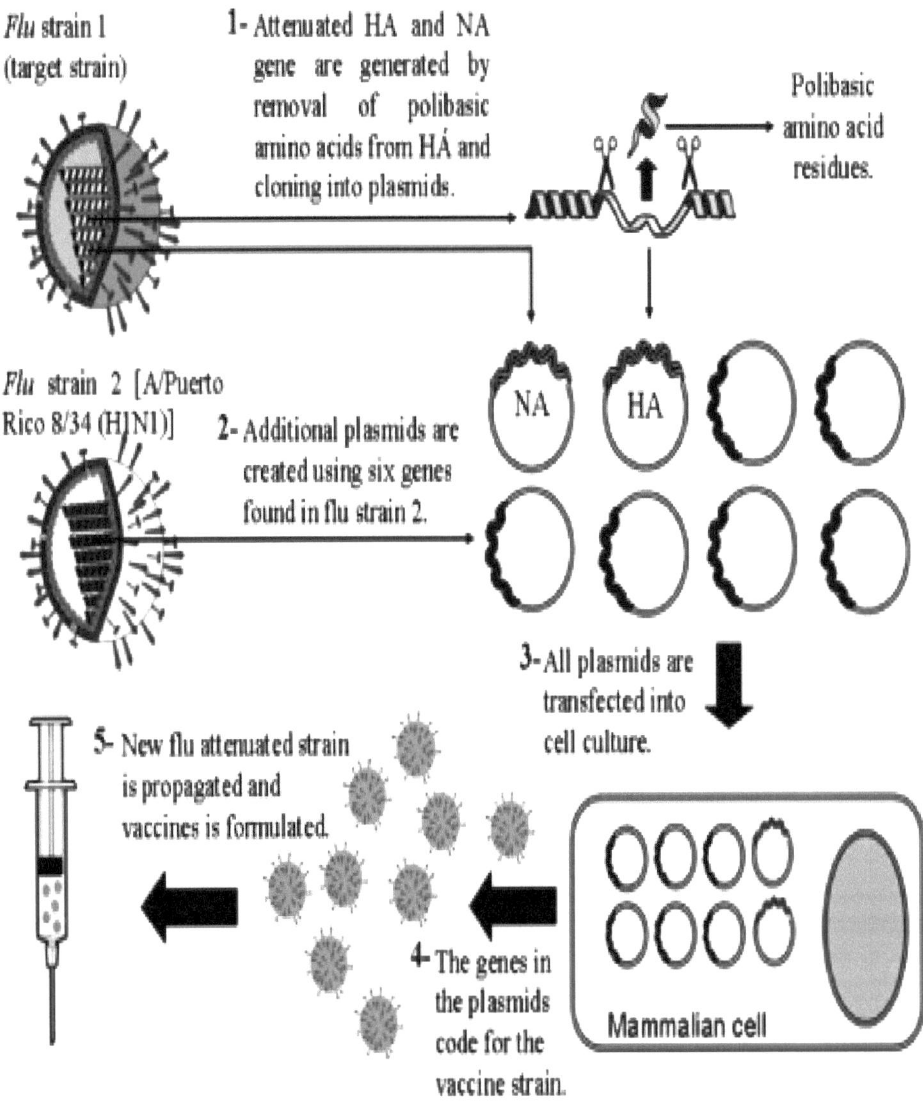

www.ingramcontent.com/pod-product-compliance
Ingram Content Group UK Ltd.
Pitfield, Milton Keynes, MK11 3LW, UK
UKHW060120300726
14090UKWH00002B/289